建筑立场系列丛书 No.70

Space for KIDS

童趣空间

MAD建筑事务所等 | 编
周一 丁树亭 孙茜 | 译

大连理工大学出版社

童趣空间

C3 建筑立场系列丛书 No.70

Space for KIDS

C3 No.70 Space for KIDS

Space Familiarizing

童趣空间

儿童教育环境设计是一项极具魅力的工作。设计师在这个过程中被引导着以孩子们的眼光去发现世界。这些环境设计是怎样激发孩子们天生的好奇心和求知欲的呢？什么样的条件才能够鼓励孩子们去探索不同的活动，从而开发个人兴趣爱好呢？在两个对立但又互补的空间品质之间达到一个平衡是至关重要的：一方面，当代幼儿园和托儿所的设计需要提供可以同时进行多种不同活动的空间，且带来丰富的视觉刺激，这样可以使孩子们参与开展自主活动。另一方面，学习空间围绕一个位于中心的、亲密的"家庭基地"设置，"家庭基地"是根据家庭规模来设计，并设有保护性的外观和入口来建立的，这样的设计为孩子们提供了熟悉且安全的环境。设计师通过这些做法给孩子们创造了一个足以模拟家庭的环境，让孩子们在这里可以自信地去探险探索周围的世界。

Designing environments for educating young children is a fascinating endeavour, because through them, they are guided in their discovery of the world. How can the design of these environments kindle children's natural curiosity and eagerness to learn? What kind of conditions encourage children to explore different activities, and to develop their personal interests? A balance between two opposing but complementary spatial qualities is needed: on the one hand, designs of contemporary kindergartens and nurseries provide spaces that can simultaneously host a variety of different activities, and that are rich in visual stimuli. In this way, they engage children in self-directed play. On the other hand, they provide a familiar and secure environment by organizing learning spaces around a central, intimate "home base", by establishing a domestic scale, and by designing a protective exterior and entrance. By doing so, they create an environment that resembles a home, from which children can confidently venture out to explore the world around them.

e for

g Adventure

ds

熟悉的探险：早期幼儿教育环境的设计

寓教于乐

儿童早年的经历对他们的发展有着至关重要的影响，因为那是他们最初认识和开始接触周围世界的时候。他们开始发展社会和语言技能，并且会受到周围环境的强烈影响。

虽然幼儿时期的生活体验在今天得到了广泛的重视，然而在此之前的一段时期并非如此。在 19 世纪中期之前，教育体系并不覆盖 7 岁以下的儿童，人们也并不认为幼儿能够学习社会经验和才智技能。[1] 直到一位德国教育学家 Friedrich Fröbel (1782—1852) 开始认识到儿童在幼年时期接受的教育对于其大脑的发展具有重要作用。[2] 他强调幼儿的"活动动力"[3] 和与生俱来的学习欲望[4]，并创立了具有教育价值的"游戏"[5]。"玩耍是人类在儿童时期发展的最高表现，"他写到，"它本是孩子灵魂的一种自由表达。"[6] 他在 1840 年建立了第一所幼儿园，孩子们在此可采用唱歌、跳舞、装饰花园和自由活动的方式来学习，Fröbel 的理论为现代学前教育奠定了基础。

世界各地的教育学者纷纷受这位德国教育学家的启发，这些追随者中还包括意大利的医学及教育学家玛利娅·蒙特梭利（1870—1952）。她开发了一种以自由活动为基础的教育方法，它需要在一个为儿童量身打造的"预设空间"中进行，此外她还提出了这种环境所应具备的空间品质。其中尤为重要的几点是，它的空间分布应当便于运动和活动，环境设计应当是优美且和谐而有序的，并且其建造比例应当适应儿童的身材。室内外皆应有自然元素的存在。[7] 虽然当时只有少数学前教育机构采纳了蒙氏教学法，但是很多幼儿园、日托中心和托儿所都具备这些空间特质。

激发好奇心的空间策略

当今，如何提供一个幼儿乐于在其中学习和玩耍的环境成为一个至关重要的话题，因为教育者面临着来自科技与广告的影响、各类标准化测试的压力以及注意力水平的普遍降低这些新的挑战。[8] 孩子们一方面需要通过一定的刺激和活动来学习；另一方面又需要一定程度的安静来培养专注力，才能消化所学习的内容。[9] 从空间的角度来说，这些对立的需求要求设计在许多方面达到一个平衡，即选址氛围必须安静，但同时具有开放和挑战性的社会化区域，建筑结构便于自由活动却又不乏有序感，面向外部环境开放却又不乏庇护感。为了回应这些挑战，现代幼儿教育环境设计也制定了一系列有助于使儿童乐于学习的空间策略。

1. 区别化的活动区

一个鼓励孩子们自由玩耍的环境要提供多种活动选择，鼓励孩子们在开发个人兴趣的过程中学习。可容纳多个活动中心的教室可以起到促进作用：一个房间的可塑性越强，功能区划分越多，其兼并容纳多样化活动的空间潜力就越大。

以 Aisaka 建筑师工作室设计的日本船桥市 Amanenomori 幼儿园为例。这处幼儿空间带有几个不同层次开放性和私密性的空间。每个房间都有一个主要活动区，一个滑动隔板可将房间连接起来，使面积增至两倍，孩子们既可在此进行团体活动又可独自玩耍。与主活动区相连的是一处天花板较低的小型空间，通往一座半开放的后花园。花园由两个育儿室共享，在与主活动区紧密连接处预留了一块相对更安静的区域。

在 Hibinosekkei+Youji no Shiro 建筑师事务所负责设计的宫古岛 Hanazono 幼儿园及托儿所项目中，建筑师们将整层地面对齐连接，再用一个个和墙等长的滑动隔板隔开，这样它就可以作为一间大教室，用于进行一些创造性的活动。在这个空间序列中，内部活动室的木质地板、木板条墙体和吸音吊顶天花板为演奏音乐和主持讲座创造了条件；画室则设置了更高一些的裸露天花板、混凝土地面和水槽，并可为一些水上活动提供场地，这些活动范围也可以延伸到半露天的庭院的木平台上。

由 Paul Le Quernec 建筑师事务所设计的法国 Niki de Saint-Phalle-Petits Cailloux 学校幼儿园的设计主旨也是容纳多种活动小组。该学校有三个中心，由三个不同直径的圆形空间重叠构成。三个空间通过楼层变化和天花板的高度来区分，两个圆形结构之间还设置了搁板及与凳子等高的储存单元。阅读室由中央带有圆顶的"豆荚"空间以及分支出来的许多大小不一的"豌豆粒"似的圆顶房间组成，同样适于不同规模的群体使用。

2. 一个带来正面刺激的环境

在儿童时期的前四年，幼儿的感知能力得到改善，他们开始开发出秩序感、方向感和社会行为。[10] 而这些方面的发展离不开一个充满正面刺激和鼓励社交互动的环境氛围的支持。

举例来说，Paul Le Quernec 设计的法国圣丹尼斯学校的外立面设计给人带来一种意在激发孩子感知能力的视觉效果。它有一个木条覆盖而成的木覆层，一面漆成橙色，一面漆成果绿色，正面保留了粗糙的表面。这样，当人们经过时，会发现建筑外观的变色效果。

西班牙加迪斯的 La Barrosa 小学附属的幼儿园由 Gabriel Verd 建筑事务所设计，公园对面的入口处突起的天篷形成了波浪的形状，加上彩色的圆柱，令这处安全的入口区域的外观看上去非常有趣。

3. 家庭基地

除了提供可划分的空间设计来容纳多元化的活动和充满正面刺激并且有利于探索的环境，创造有利于学习的环境往往还需要有一个中央空间。它应该是半封闭的，向上而非向外开放，内部氛围温暖而熟悉。它应具有高辨识度，定位应该是"家庭基地"或者欢迎孩子们归来的温暖"巢穴"。[11]

studio we 事务所设计的瑞士斯塔比奥的一所幼儿园的中央空间设有橙色的巨大天窗和三面墙壁，明显与学校内部其他白色的开放区域区分开来。圆形的天窗带来一种静态的、中心的空间定位，并从天花板垂下来，形成一个保护区。

在 Dominique Coulon & Associés 主持设计的位于法国 Buhl 的一所幼儿园项目中，作为核心的"家庭基地"相比之下就要大得多，它同时也被用作入口门厅，安置了进入周围小型房间的通道。但如此大的规模并不影响它带给孩子们应有的私密感觉。天花板的几何形状和地板的材质将空间分隔成四个区域，沿着墙面安置的储物柜单元同时也为孩子们提供了座位。双高墙体的地面和较低部分墙面均为粉色和紫色系的暖色调。高高的窗户设计提供了充足的自然采光，赋予建筑一种明亮且热情的氛围。

4. 家庭规模

作为家庭和小学之间的一个过渡，儿童早期教育机构还可以通过模拟家庭的规模来营造熟悉的环境氛围。

MAD 建筑事务所的日本爱知县四叶草幼儿园项目正是以家庭规模为主要设计特点。设计师将当地幼儿园园长的私人家庭住房进行整修，白天将其作为一个适合孩子全面发展的教育机构，晚上仍然保留其住宅功能。房子的木质结构覆以白色圆形的洞状外壳。内部房间的比例、倾斜的屋顶以及风化的木质房梁使这所幼儿园成为一个像家一样的地方。

然而，Bruno Fioretti Marquez 事务所设计的瑞士卢加诺的幼儿园及托儿所项目则占据了较大的区域面积，该建筑具有家庭住宅的模式，拥有 56 个类似于住宅的单元，镶嵌在 7m×8m 的网格框架中。每个班级由五个这样的单元组成。在这些教室之间，有些住宅单元是外部庭院，其他一些则连接在一起形成一个中央覆盖的街道。不同朝向的屋顶的景观遵循着房子的形状，使整个幼儿园看起来像是一个小村庄。

Pierre-Alain Dupraz 事务所的瑞士普兰京斯幼儿园项目也遵循了住宅规模的设计。四个方形建筑体量交叉排列成十字形，每个单元的楼层高度递增三分之一。每个建筑体量内部容纳两间教室以及带坡道、楼梯和双层高度的"可坐下休息的阶梯"的流线区。这些流线区在中心部位重叠在一起，形成了由阶梯、座位区以及与教室衔接的平台参差交错的景观。

5. 保护性的外部环境

根据所处的环境和儿童的需求特点，一些育儿机构的室外和出入口设计旨在给人以坚固安全的感觉。这样的建筑通常使用夸张的几何形状和坚实材料装饰的坚固墙壁，让人感觉进入了一处被严密保护的区域。

Taller Básico de 建筑事务所主持设计的西班牙 Haro 幼儿园的夸张的棱角形状和粗糙的混凝土框架，形成了一种强有力的保护氛围，使学校外观看上去像是坐落在山顶的一块巨石。天花板向下倾斜，墙壁向入口靠拢汇合。一但进入室内，孩子们立即进入一个全玻璃的、面向外部开放的、带向上倾斜的天花板的教室，从一个安全的位置向孩子们展示周围的景象。

对于澳大利亚汤斯维尔的受虐儿童信托中心来说，安全感是尤为重要的设计主旨。该项目由 m3architecture 事务所主持设计，整

Cassarate幼儿园，瑞士
Cassarate Kindergarten, Switzerland

La Barrosa幼儿园，西班牙
La Barrosa Infant Pavilion, Spain

体由两座坚固的保护性入口建筑组成，两座楼之间设有封闭式花园和治疗室。外部建筑带有拱形窄小窗户，让人联想到城堡或堡垒般的墙壁。

幼儿园及托儿所通过提供两种对立性质的环境来鼓励儿童参与到游戏和学习中：一方面，空间提供各种游戏活动区，有趣的建筑元素给孩子们提供了自主玩耍和探索的机会；另一方面，以中央家庭基地、家庭规模以及保护性的室外为特色的环境给孩子们带来熟悉和安全的感觉。文中提到的这些设计策略想表达的是，只有当孩子们身处一个安全可靠的环境中——带给他们家的感觉——他们才能够带着好奇心去探索周围世界。

Familiarizing Adventure: Designing Environments for Early Childhood Education

Learning Through Play

Children's experiences during their early years have a great impact on their development, as it is during this time that they are introduced to the world around them, and begin to engage with it. They begin to develop social and language skills, and are strongly affected by their environment.

While the importance of early childhood experiences is widely recognized today, however, it was not always so. Before the mid-19th century, there was no educational system for children younger than seven years of age, nor was it recognized that young children were able to learn social and intellectual skills at all.[1] It was not until Friedrich Fröbel (1782-1852), a German educator, recognized the significant brain development that occurs during a child's early years that any importance was given to early-years education.[2] He emphasized young children's "activity drive"[3] and innate desire to learn[4], and established the educational worth of the "game" and of play.[5] "Play is the highest expression of human development in childhood," he wrote, "for it alone is the free expression of what is in a child's soul."[6] By establishing the first kindergarten in 1840, where children learned through singing, dancing, gardening, and self-directed play, Fröbel paved the way for modern-day pre-school education.

Educators world-wide were inspired by the German educator's approach, among which was the Italian physician and educator Maria Montessori (1870-1952). In addition to developing a teaching method based on free activity within a "prepared environment" tailored to the children's needs, she proposed spatial qualities that such an environment should have. In particular, it should be arranged in a way that facilitates movement and activity; it should be beautiful, harmonious and orderly, and constructed in proportion to the child. As well, there should be nature both inside and outside the classroom.[7] While only a small proportion of pre-school education follows the Montessori teaching method, many kindergartens, daycares and nurseries have these spatial characteristics.

Spatial Strategies for Engaging Curiosity

Today, the question of how to provide environments that engage young children in learning and play is a critically important topic, as educators face new challenges relating to the influence of technology and advertising, pressures

1. Miriam LeBlanc, "Friedrich Froebel: His life and influence on education", Community Playthings, http://www.communityplaythings.co.uk/learning-library/articles/friedrich-froebel
2. "Brief History of the Kindergarten", Froebel Gifts, http://froebelgifts.com/history.htm
3. Friedrich Froebel, "The Education of Children", translated by J.Liebschner, 1844 http://www.friedrichfroebel.com/
4. Miriam LeBlanc, "Friedrich Froebel: His life and influence on education", Community Playthings
5. Stanley James Curtis, "Friedrich Wilhelm August Fröbel", Encyclopaedia Britannica, https://www.britannica.com/biography/Friedrich-Froebel
6. Miriam LeBlanc, "Friedrich Froebel: His life and influence on education", Community Playthings
7. Christine Ann Christle, "Montessori School", Encyclopaedia Britannica, 2016.5.11, https://www.britannica.com/topic/Montessori-schools
8. Miriam LeBlanc, "Friedrich Froebel: His life and influence on education", Community Playthings
9. *The Adventure of Question and Answer: Proposal for a new school*, s-Hertogenbosch: Magnolia Foundation, May 2016, p17
10. Christine Ann Christle, "Montessori Schools"
11. Herman Herzberger, Space and Learning, Rotterdam: 010 Publishers, 200, p35

of standardized testing, and diminished levels of concentration.[8] On the one hand, children need stimulation and activity; on the other hand, they require quietude to develop concentration, and to process what they learn.[9] Spatially, these opposing needs require a balance between quiet areas and open, stimulating social zones, between freedom to move around and a structure that gives a sense of order, and between openness to the external environment and a sense of shelter and protection. In response to such challenges, contemporary designs for childhood education environments demonstrate a number of spatial strategies that engage children in learning:

1. Differentiated activity zones

An environment that fosters self-directed play provides children with a variety of activities to choose from, encouraging them to learn by developing their personal interests. It is facilitated by classrooms that allow for multiple centers of focus: the more a room is moulded or differentiated into various zones, the more spatial potential it has to simultaneously host a diversity of learning activities.

For instance, Aisaka Architects' Atelier has designed nursery rooms in the Amanenomori Nursery School in Chiba, Japan with several zones with varying levels of openness and intimacy. Each room has a main activity zone that can be doubled in size by connecting it to an adjoining nursery room via a sliding partition, allowing groups to work together or separately. Connected to this main area is a smaller space with a lowered ceiling, which gives access to a semi-enclosed back garden. This garden is shared by two nursery rooms, and offers a quieter area that is still closely linked to the main activity space.

Hibinosekkei + Youji no Shiro architects have aligned the ground floor spaces of the Hanazono kindergarten and nursery in Miyakojima, Japan, and connected them with wall-length movable partitions, so that they can function as one large classroom for creative activities. Within this sequence of spaces, the studio's wood floor, wood batten wall and acoustic ceiling create a setting for playing music and hosting lectures; the atelier, with a higher, exposed ceiling, concrete floor and sink provides a place for wet activities; these activities can spill out onto the wood deck of the partially covered courtyard.

The nursery rooms in Paul Le Quernec's "Niki de Saint-Phalle – Petits Cailloux" school in Saint Denis, France, are also designed to accommodate multiple activity groups. They each have three centers, being composed of three overlapping circular spaces of different diameters. The three spaces are differentiated by means of level changes and ceilings, with shelves and seat-height storage elements between two circles. As well, the reading area is composed of a central domed room with smaller domed "pods" branching off from it, lending itself to activities for both larger and smaller groups.

2. A stimulating environment

During their first four years, children's senses are being refined, and they develop a sense of order, orientation, and

照片提供：©Wison Tungthunya

Jerry住宅，泰国
Jerry House, Thailand

social behaviour.[10] The development of these aspects is supported by an environment that is rich in positive stimuli and that encourages social interaction.
The facade of Paul Le Quernec's School in Saint-Denis, for instance, is designed with an effect of optical illusion that is meant to stimulate children's sense of perception. It is a wooden cladding system with battens that are painted orange on one side and apple green on the other, with the front facets left rough. This way, the facade changes color as one passes by.
In the Infant Pavilion for the Primary School La Barrosa in Cadiz, Spain, designed by Gabriel Verd, the canopy bulges out near the entrance facing the park, and its undulating shape and colorful circular columns give this protected entrance area a playful appearance.

3. A home base

In addition to providing a differentiated space that can accommodate diverse activities, and a rich environment that stimulates and provides opportunities for discovery, learning environments also often have a central space that is partially enclosed, oriented upwards instead of towards the outside, and that has a warm and familiar atmosphere. It is readily identifiable, and lends itself to functioning as a kind of "home base" or "nest" that children can return to.[11]
In studio we's design for a Kindergarten in Stabio, Switzerland, this central space has a large, orange-coloured skylight and walls on three sides, separating it from an otherwise open and white school interior. The skylight's circular shape gives a static and centered spatial orientation, and drops down from the ceiling, creating a sheltered zone.
The central "home base" at the Nursery in Buhl, France, designed by Dominique Coulon & Associés, is much larger in comparison, and also functions as an entrance hall, providing access to the smaller rooms arranged around it. Despite its scale, however, it maintains a sense of intimacy. The ceiling geometry and floor materials divide the space into four quadrants, and wardrobe elements along the walls provide nooks where children can sit. As well, the floor and lower portion of the double-height walls are coloured with a spectrum of warm pink and purple tones. High windows provide abundant natural light, giving it a bright and welcoming atmosphere.

4. Domestic scale

As a place of transition between the home and primary school, early childhood learning environments create a familiar environment by having a domestic scale.
This familiar scale characterizes the Clover House Kindergarten in Aichi, Japan, for instance. MAD Architects have renovated the local kindergarten owner's family house to accommodate a fully developed education institution by day, and retain its residential function at night. The house's wood structure is covered with a white, rounded cave-like shell. Inside, the proportions of the rooms, the house's pitched roof and the weathered wood beams make the kindergarten into a place that feels like a home.
While the Kindergarten and Nursery in Lugano, Switzerland by Bruno Fioretti Marquez occupies a larger area, it maintains a domestic scale with 56 house-like units that are tesselated in a 7x8 grid. Each class is composed of five such

照片提供：©Peter Bennetts

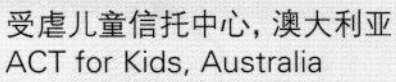
受虐儿童信托中心，澳大利亚
ACT for Kids, Australia

照片提供：©Pedro Pegenaute

Haro幼儿园，西班牙
Nursery School in Haro, Spain

units. Between the classes, some of the house-units are exterior courtyards, while others are joined together to form a central covered street. As well, the landscape of differently oriented shed roofs makes a formal reference to the house shape, giving the kindergarten an appearance of a small village.

Pierre-Alain Dupraz's Prangins Kindergarten in Switzerland, also creates a domestic scale. Four rectangular volumes are arranged in a cross-shaped plan, and step up by a third of each unit's floor height. Each volume contains two classrooms and circulation zone with ramps, stairs and double-height "sitting stairs". These circulation zones overlap in the core, resulting in a cascading landscape of stairs, sitting places and landings that connect the classrooms.

5. A protective exterior

Depending on the context and on children's needs, some childcare building's exteriors and entrances are designed to give a sense of solidity, safety and security. They do so by using solid walls with bold geometries and heavy materials, giving an experience of entering a protected place.

For instance, the bold, angular geometry and rough concrete formwork of the Nursery School in Haro, Spain by Taller Básico de Arquitectura creates a sense of strength and protection, and gives the school the appearance of a heavy rock that sits on its hilltop location. The ceiling slopes down and the walls converge towards the entrance. Once inside, however, children enter classrooms that are completely glazed towards the outside, with a ceiling that slopes upwards, opening up the view to the surrounding landscape from a secure place.

Creating a sense of security is particularly important for ACT for Kids in Townsville, Australia. m3architecture has designed a complex with two solid, protective entrance buildings, with enclosed gardens and therapy rooms between them. The outer buildings are materialized with deep, narrow arched windows, evoking castle- or fortress-like walls.

Kindergartens and nurseries engage children in play and learning by providing environments that have two opposing qualities; on one hand, spaces offer a variety of playful activity zones, and intriguing building elements give children opportunities for self-directed play and discovery. On the other hand, environments with a central home base, a domestic scale, and a protective exterior give a sense of familiarity and security. These design strategies are a reminder that it is only when children are in a secure and trusted environment – when they feel at home – that they can engage their curiosity in discovering the world around them. Isabel Potworowski

La Barrosa幼儿园

Gabriel Verd Arquitectos

基于霍华德·加德纳的多元智力理论和蒙特梭利学校的教学法理念，我们设计了一座全新的幼儿园。这座新幼儿园拥有有机的弧线和流畅的线条，并与景观相融合。

La Barrosa 幼儿园内的孩子都是 5 岁，该学校急需一处新空间，来满足 3～5 岁孩子的要求。因此，我们在周围约 4000m² 的场地进行了扩建，该场地高低不平，高度差为 5m，建有教室和活动场所，来与原有的幼儿园相互补充。

从经济的角度来看，我们设计了一个紧凑的体量，教室的西立面或东立面能够在清晨和中午直接获得日光。

植物的分布也与简单的功能布局相呼应，1000m² 的空间可容纳 225 名儿童。十间教室沿着走廊分布，走廊两侧分布着高窗，高窗为走廊提供照明，同时还起到通风的作用。在这一层，维修人员还可通过隐蔽的内室来对人工照明设备和其他设施进行简单的维修。

室外的投影区域对操场提供了保护，使其免受盛行强风的影响，同时也给孩子们提供了必要的隐私性，使其在玩耍时不被街上的人们看到。

建筑外侧的柱子林立，支撑着蜿蜒的混凝土门廊。柱子和门廊形似树干和树冠，对幼儿园提供了保护。绿色的瓷砖使人们想起了不久之前还耸立在这里的松木，并且赋予整体建筑与众不同的形象。

La Barrosa Infant Pavilion

We design a new Infant Pavilion based on the principles of Multiple Intelligences of Howard Gardner and pedagogical methodology of the Montessori schools. A new building with organic curves and smoothed outlines integrates into the landscape.

With just five years old children, the "La Barrosa" school needs a new space for satisfying the demand of younger students between 3 and 5 years old. We had an adjacent plot of 4,000 sqm with an unevenness of 5 meters to place the classrooms and sports spaces that implement the existing ones.

In an economical thinking, we have designed a compact

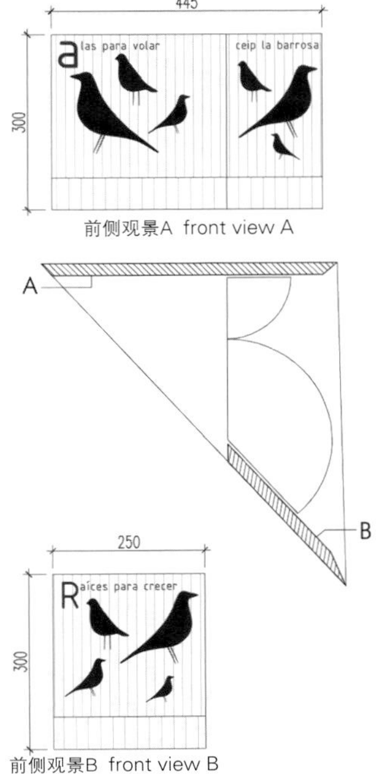

前侧观景A front view A

前侧观景B front view B

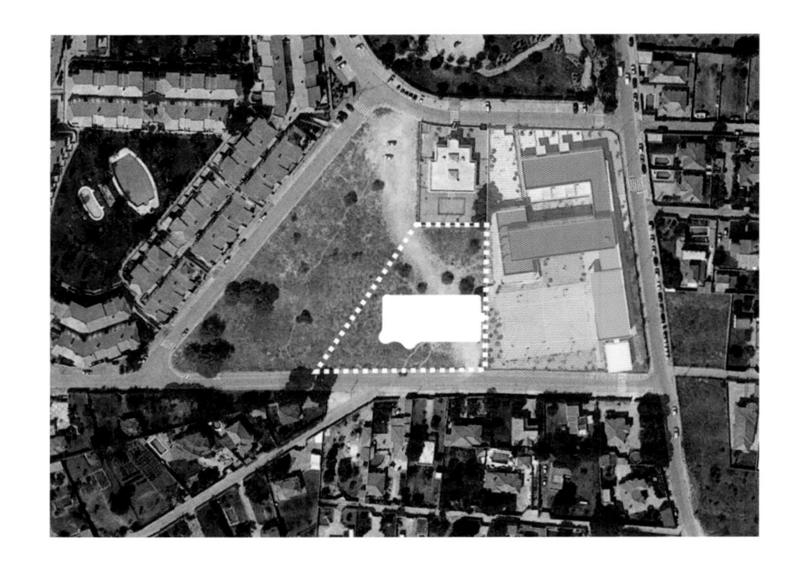

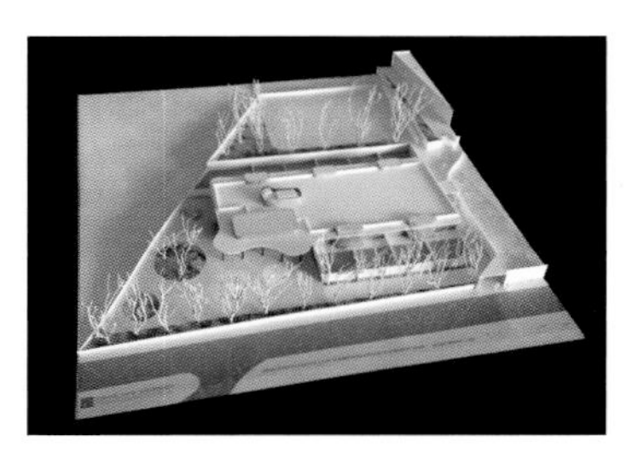

1.主入口 2.教室 3.多功能教室 4.室外教室 5.教师办公室 6.存储间 7.浴室 8.设备间&消防控制室
9.老学校 10.操场-浴室 11.体育场 12.操场 13.与老学校连接的地方

1. main entrance 2. classroom 3. multipurpose classroom 4. outdoor classroom 5. teacher's room 6. storage 7. bathroom
8. installations & fire prevention room 9. old school 10. playground-bathroom 11. sports field 12. playground 13. connection to the old school

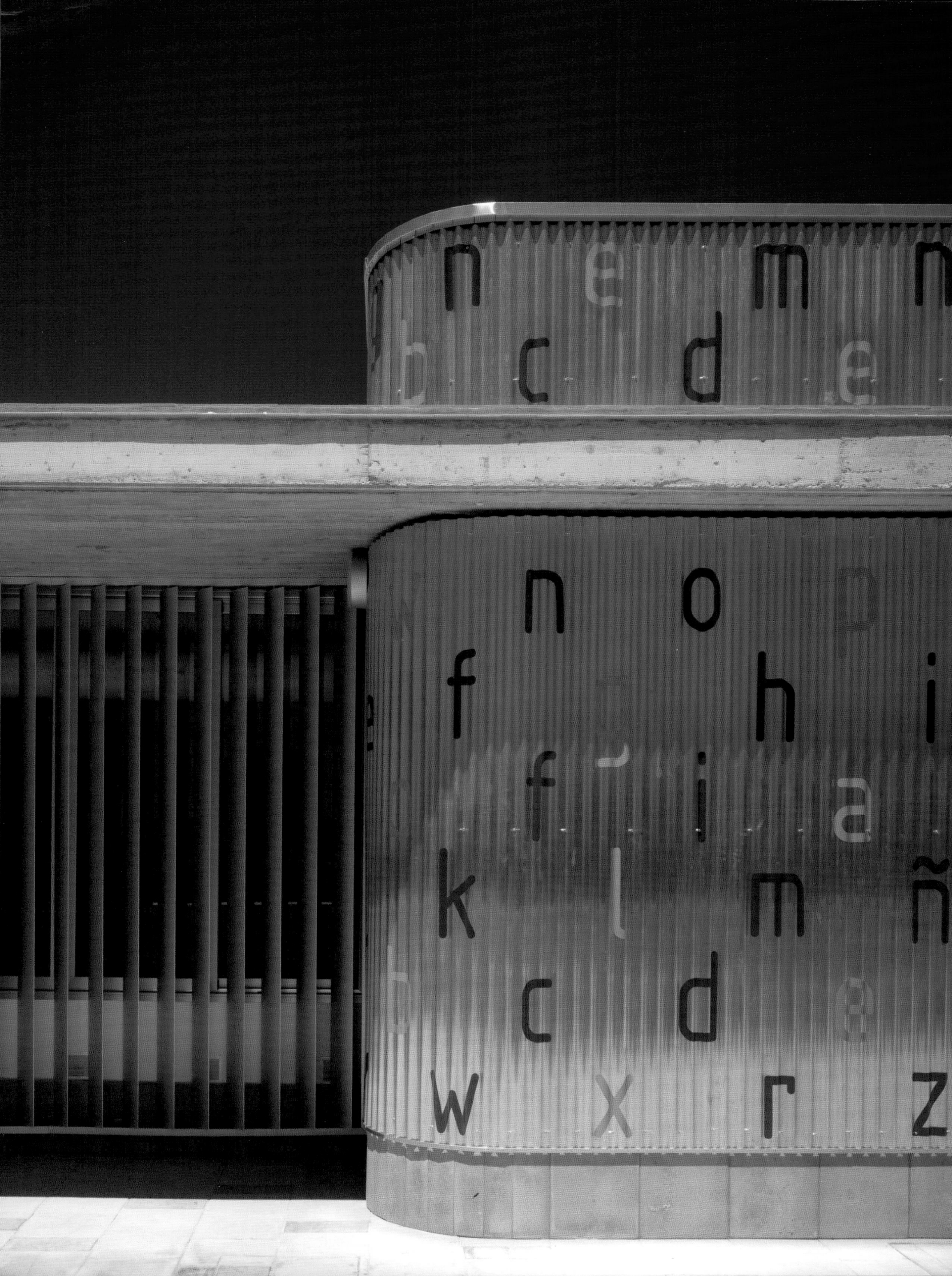

volume. The classrooms have been distributing in order to get direct sunlight, either early morning or noon (east or west facade)

The configuration of the plant responds to a simple layout for a simple program, with 1,000 sqm to accommodate 225 children. The ten classrooms are distributed through a corridor which is illuminated and ventilated by high windows on both sides. In this level the artificial lighting and facilities are accessible through hidden trays for an easy maintenance.

The projected section protects playgrounds from prevailing strong winds and provides the necessary privacy for children to play hidden from street views.

The small forest of pillars that hold the concrete porch with sinuous shapes was conceived as an allegory to the trunks and treetops that protect the first schools of our civilization.

The green tones of tiles recall the pinewoods that not long ago occupied the area and give the whole its distinctive image. Gabriel Verd

GRAY 1 KERAMEX	WHITE 5	[illegible] 2	IVORY 3 KERAMEX	GREEN WATER 4 KERAMEX	[illegible] 2	IVORY 3 KERAMEX	WHITE 5	GRAY 1 KERAMEX	GREEN WATER 4 KERAMEX
IVORY 3 KERAMEX	GREEN WATER 4 KERAMEX	WHITE 5	GRAY 1 KERAMEX	[illegible] 2	WHITE 5	GRAY 1 KERAMEX	GREEN WATER 4 KERAMEX	[illegible] 2 KERAMEX	IVORY 3 KERAMEX
GRAY 1 KERAMEX	[illegible] 2	IVORY 3 KERAMEX	GREEN WATER 4 KERAMEX	WHITE 5	[illegible] 2	IVORY 3 KERAMEX	GRAY 1 KERAMEX	WHITE 5	GREEN WATER 4 KERAMEX
GREEN WATER 4 KERAMEX	WHITE 5	GRAY 1 KERAMEX	[illegible] 2	IVORY 3 KERAMEX	WHITE 5	GRAY 1 KERAMEX	GREEN WATER 4 KERAMEX	[illegible] 2 KERAMEX	IVORY 3 KERAMEX
IVORY 3 KERAMEX	GREEN WATER 4 KERAMEX	[illegible] 2	WHITE 5	GRAY 1 KERAMEX	IVORY 3 KERAMEX	WHITE 5	GRAY 1 KERAMEX	GREEN WATER 4 KERAMEX	[illegible] 2
PISTACHO 2	GRAY 1 KERAMEX	IVORY 3 KERAMEX	GREEN WATER 4 KERAMEX	WHITE 5	GRAY 1 KERAMEX	GREEN WATER 4 KERAMEX	[illegible] 2	IVORY 3 KERAMEX	WHITE 5

照片提供：©Álvaro Rodríguez (courtesy of the architect)

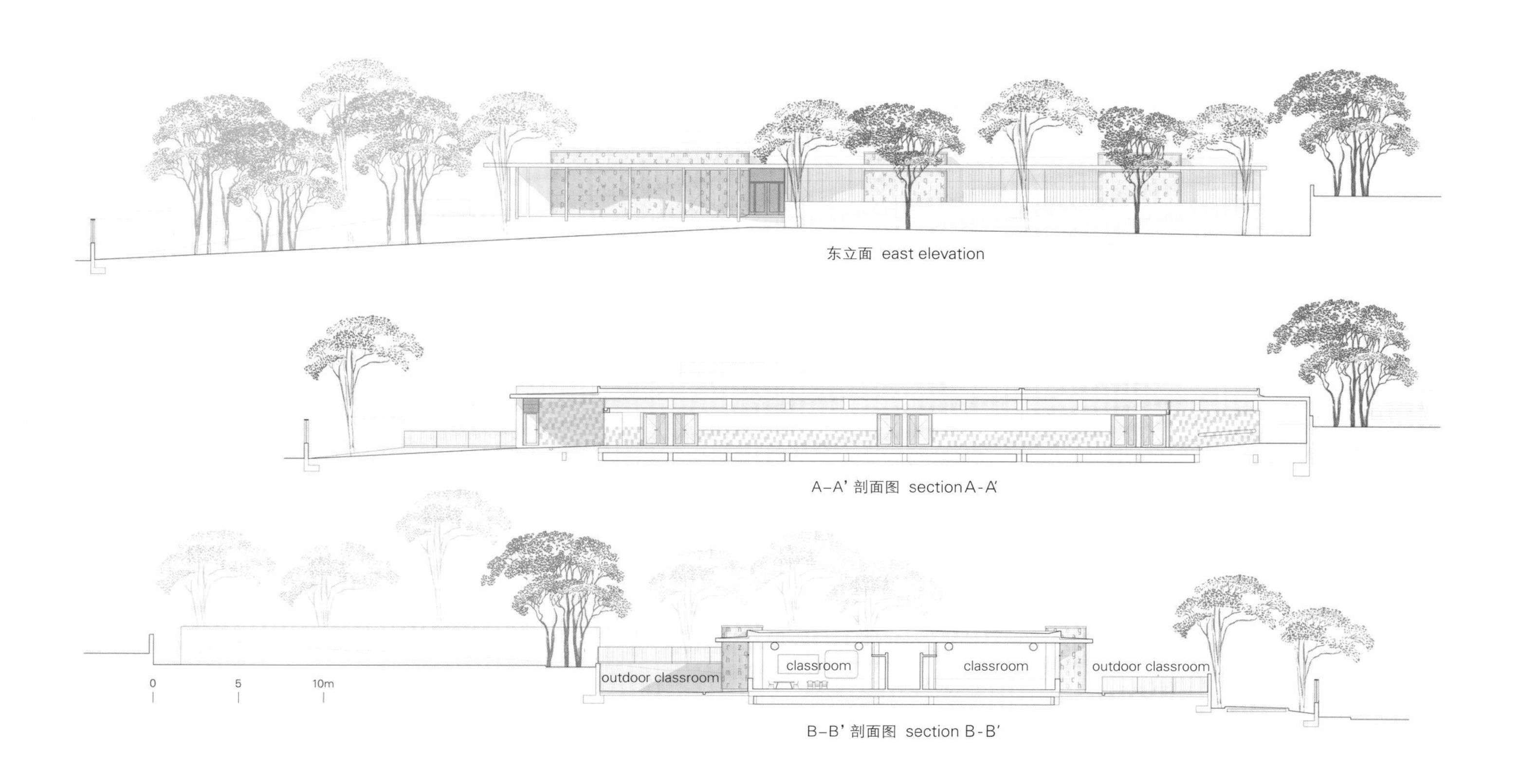

东立面 east elevation

A–A' 剖面图 section A-A'

B–B' 剖面图 section B-B'

项目名称：New infant pavilion for the public primary school “La Barrosa”
地点：Chiclana de la Frontera, Cadiz
建筑师：Gabriel Verd
合作：Eduardo Vazquez, Daniel Yusty Lopez_Technical architect,
Pablo Portillo_Building services, Elías Pérez de Lema_Structural engineers,
Elisabetta Mosti_Architect, Fiametta Conforti y Santina Tegas_Students
客户：Consejería de Educacion de la Junta de Andalucía / Infraestructuras y Servicios Educativos (ISE)
施工方：Diaz Cubero S.A.
总建筑面积：1,000m^2 / 造价：EUR 1,213.898
竞标时间：2013.5 / 竣工时间：2015.8
摄影师：©Jesus Granada (courtesy of the architect) (except as noted)

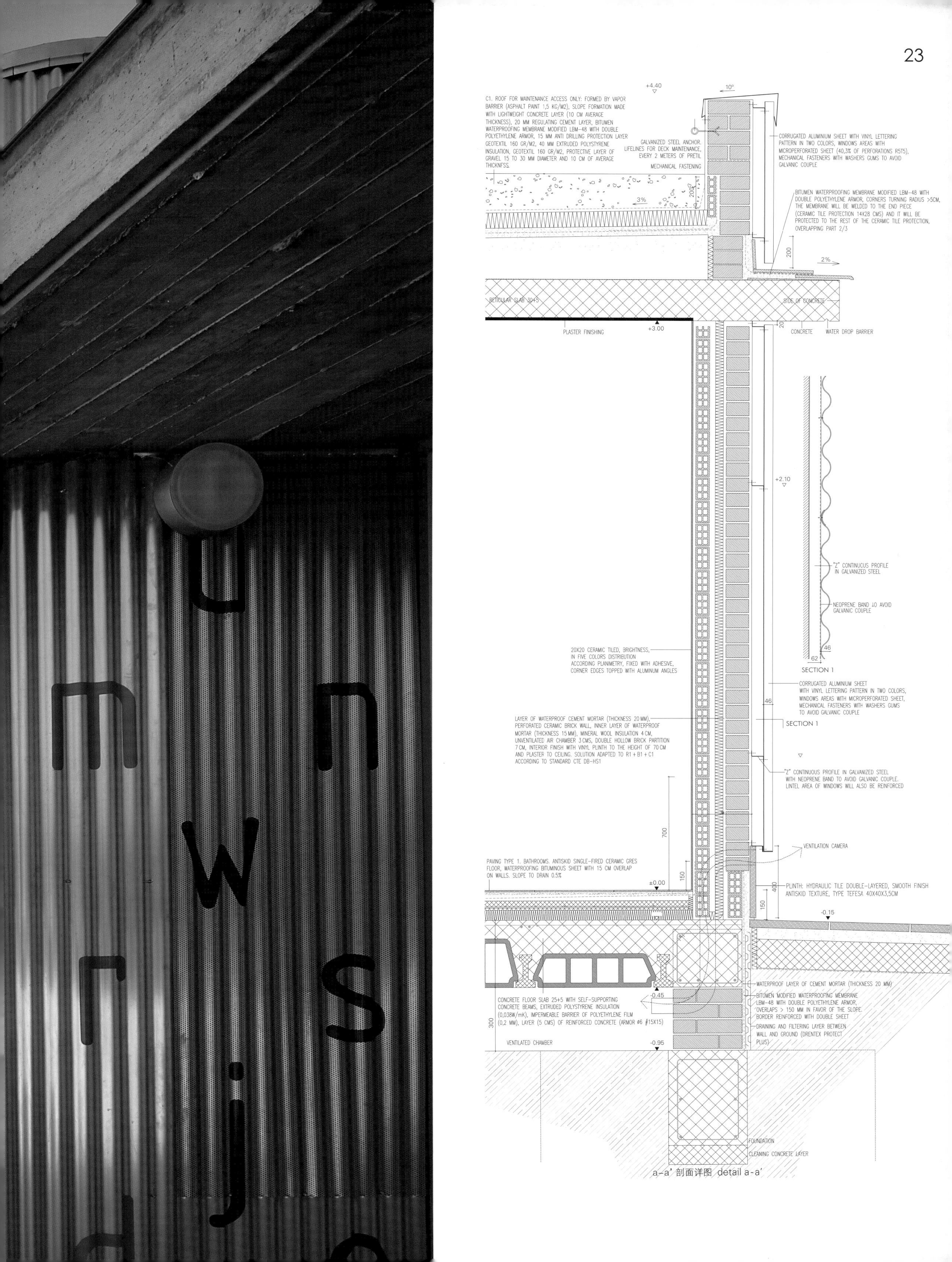

a-a' 剖面详图 detail a-a'

受虐儿童信托中心

m3architecture

这个项目为受虐待的儿童提供了一处如堡垒般的家园。

位于汤斯维尔的詹姆斯·库克大学的受虐儿童信托中心（ACT）面向校园内 20 世纪 60 年代建造的建筑师詹姆斯·博雷尔基金会大楼。部分建筑采用混凝土和灰面混凝土砌砖建造，造型简单，形式可塑。

在公共领域，这些材料具有非常突出的特点——坚固，且能够使人产生愉悦感。对于新建筑来说，这些都是非常合适的材料，其内可容纳行政区、日托中心，以及最重要的，一系列治疗区域。

该建筑的外围由混凝土建造，室内密室为治疗区，两片高大的树林围合出房间的形状，且和经过验证给治疗带来益处的景观相呼应。

我们的新项目要求这处公共领域要真正应儿童的需求而建，不仅仅为每一个儿童，尤其还要为那些有着极端经历、急需入住治疗区的孩子建造设施。

这座建筑利用进一步简化的几何外形和建造方法，在坚固性和令人愉悦的功能之间达到了平衡。

ACT for Kids

This project acts as a strong, fort-like home for abused children.

The Abused Children's Trust (ACT) for Kids project at James Cook University in Townsville looked to the context of architect James Birrell's late 1960s foundation buildings on the campus. Some of these are in concrete and grey face concrete blockwork, with simple geometries and plasticised forms.

In the public realm, these are fine qualities – firmness with a sense of delight. These were thought to be suitable materials for the new building accommodating administration, day care and most importantly, a suite of therapy spaces.

一层 ground floor

二层 first floor

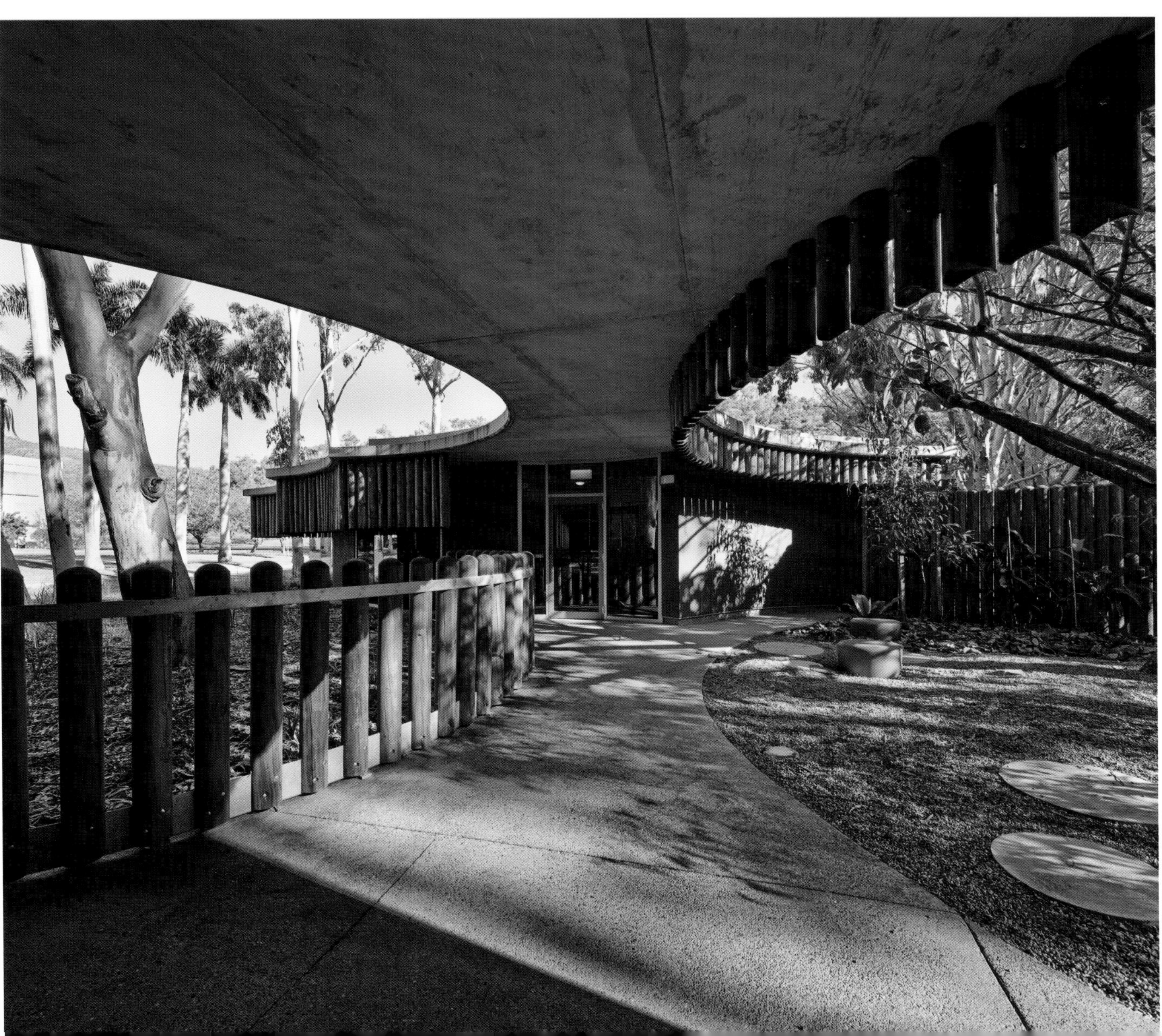

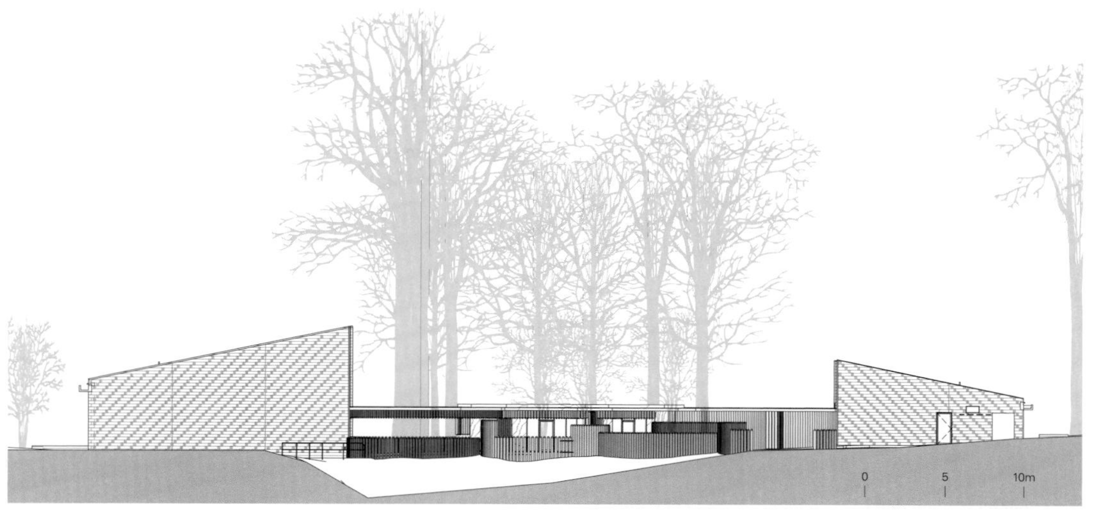

西北立面 north-west elevation

项目名称：ACT for Kids / 地点：Townsville, Australia
建筑师：m3architecture
结构工程师：Bligh Tanner / 工料测量师：Steele Wrobel
本地建筑师：Stephen de Jersey / 景观建筑师：Lat 27
承包商：Hutchinson Builders / 客户：ACT for Kids
功能：administration, therapy and day care
用地面积：2,400m² / 总建筑面积：850m²
结构：concrete/concrete blockwork
室外饰面材料：concrete blockwork / 室内饰面材料：plasterboard
造价：USD 4.5M / 设计时间：2011 / 竣工时间：2014
摄影师：©Peter Bennetts (courtesy of the architect)

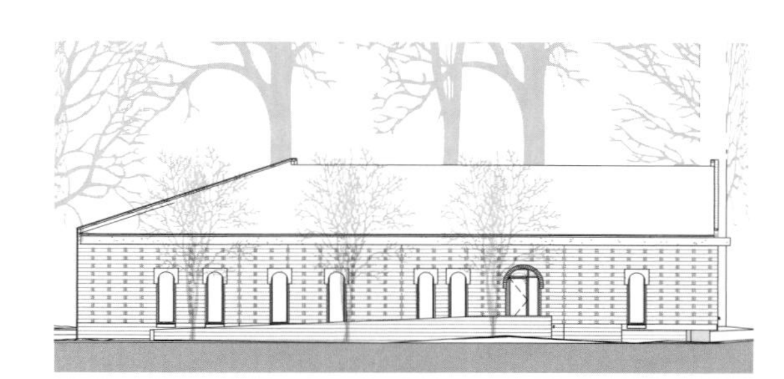

东北立面 north-east elevation

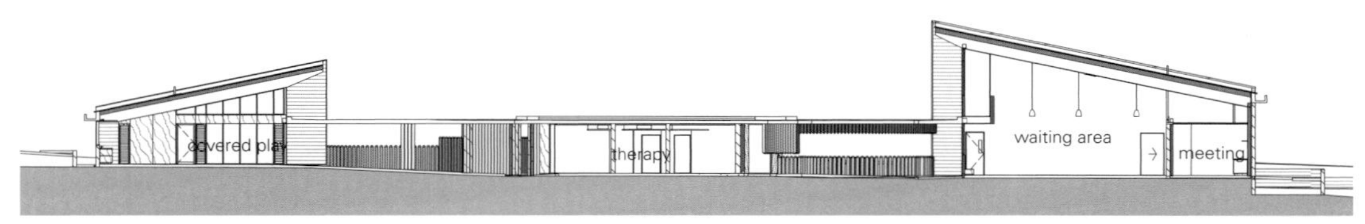

A–A' 剖面图 section A-A'

The design has a concrete block perimeter, with therapy rooms in the inner sanctum, sculpted around two stands of trees – in response to the well-documented benefit that landscape brings to therapy.

Our new project asks for what a public realm agenda for children might be like – and not just any children, but those who are in extreme circumstances so as to require this facility. With further simplified geometries and means, the building provided a balance of firmness and delight.

粗犷游乐场

Assemble + Simon Terrill

这个项目是英国皇家建筑协会公共画廊中的一个设施。它对三处富有特色的英国住宅区中的三个分支结构进行了 1:1 的复制，分别为皮姆利科区的丘吉尔花园、波普拉区的棕色地带以及帕丁顿区的布鲁内尔住宅。Assemble 建筑事务所将这些混凝土和钢材质的游乐场结构进行了重塑，利用再生泡沫材料在新场地建造，使人们摒弃从材料的角度来考虑其形式特点。此外，它们作为游乐场地，还重新得到了评估。

游客可以爬上淡粉色、蓝色以及绿色的物体之上，这些颜色的物体构成了楼梯、斜坡、平台、滑道，以及直径为 5m 的圆盘场地，场地设有黄色的金属栏杆，一侧还被抬高了设置。

观景平台向上延伸至天花板处，人们可在此透过画廊的玻璃屋顶向外观看风景。这个观景平台再现了（棕色地带）Balfron 塔楼基座内游乐场内的部分结构。该塔楼由 Erno Goldfinger 设计，是仍位于原场址的仅有的几个原始设计之一—Assemble 和 Simon Terrill 利用泡沫材料（泡沫板是一种应用在室内地面的常见材料）对这些外形进行了复制，意图向人们征询鼓励孩子玩耍的方式。

Brutalist Playground

This project is an installation for the RIBA public gallery. It consisted of full size fragments of three distinctive London housing estates: Churchill Gardens in Pimlico, the Brownfield Estate in Poplar, and the Brunel Estate in Paddington. Assem-

照片提供：©John Maltby

皮姆利科区的丘吉尔花园，伦敦，1962年
Churchill Gardens, Pimlico, London, 1962

照片提供：©Assemble and Simon Terrill

波普拉区棕色地带的Balfron塔楼游乐场，伦敦，1967年
Balfron Tower playground/Brownfield Estate, Poplar, London, 1967

照片提供：©John Donat

帕丁顿区的布鲁内尔住宅，伦敦，1962年
Brunel Estate, Paddington, London, 1962

项目名称：The Brutalist Playground
地点：Royal British Institute of British Architects public gallery
建筑师：Assemble
项目团队：Joseph Halligan, Jane Hall
合作：Simon Terrill
结构工程师：Structure Workshop, Flux Metal
施工方：Isabel + Helen
材料：Reconstituted foam(chip foam)_sourced from Custom Foams
竣工时间：2015.8.16
摄影师：
©Tristan Fewings (courtesy of the architect) (except as noted)

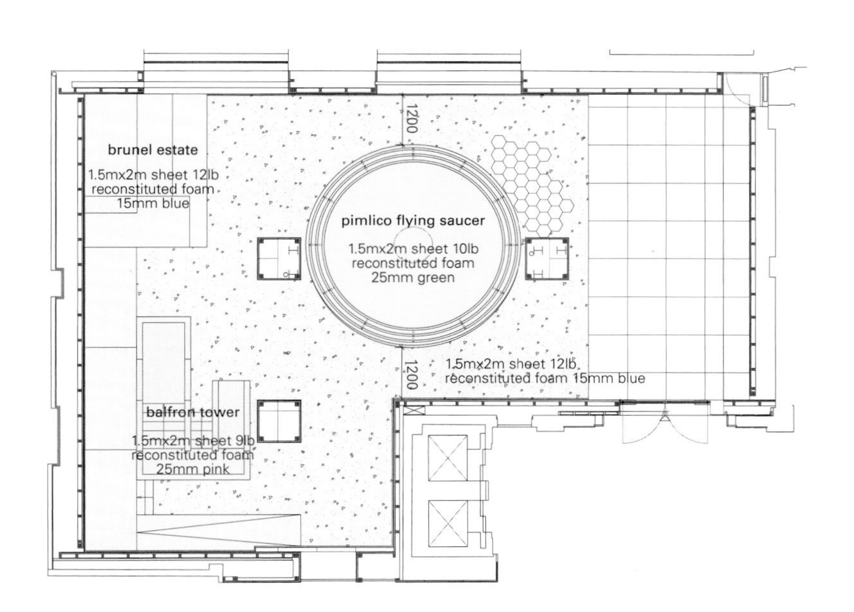

ble recast these concrete and steel playground structures in reconstituted foam in order to allow people to consider their formal characteristics separately from their materiality, and in doing so allow them to be reappraised as places for play. Visitors can climb on the pastel pink, blue and green objects, which form stairs, slopes, platforms and a slide, as well as a five-meter-wide disc with a yellow metal balustrade that is elevated at one end.

A lookout platform that extends through the ceiling, offering views out of the gallery's glass roof, is a recreation of part of the playground at the base of the Balfron Tower, designed by Erno Goldfinger – one of the only original designs still in situ. By replicating the shapes in foam – a material commonly found in indoor playgrounds – Assemble and Simon Terrill also intended to raise questions about the way children are encouraged to play.

四叶草幼儿园

MAD Architects

MAD 建筑事务所完成了其在日本设计的第一个项目，即四叶草幼儿园。该项目位于日本冈崎的一座小镇中，孩子们在幼儿园内可看见稻田和山丘的美景，而这也是爱知县的特色。

这座幼儿园是由 Kentaro Nara 和 Tamaki Nara 兄妹的老房子改造而成的。不久之后，这座老房子就因为太小而无法满足扩招的需求。因此，这对兄妹希望建造一座现代化的教育设施，使孩子们在这里能感觉到如家般的舒适，在一处充满了教育氛围的环境中成长和学习。

MAD 建筑事务所接受了委托，将业主的老式二层住宅改造为一座全面开发的教育设施。改造过程从对现有的 105m² 的住宅的调查开始。这座木构住宅和周围的住宅一样，是一座标准的预制建筑。为了将施工成本降至最低，MAD 建筑事务所决定回收原有的木结构，并将其融入新设计中。原有的木结构都应用在主要的教学区，使之成为四叶草幼儿园历史的象征性回忆。其半透明的封闭空间可轻易地举办不同的教学活动。窗户的外形各不相同，孩子们非常易于识别，阳光可透过窗户洒进室内，创造了不断变化的阴影区，由此激发孩子们的好奇心，鼓励他们发挥想象力。

这座新幼儿园的表皮和结构将原有的木结构包裹起来，如同一件外衣，将建筑的骨架覆盖，在新旧结构之间创造了一处模糊的空间。四叶草幼儿园的设计起点是其富有标志性的斜屋顶。改造中使用的构件营造了动态的室内空间，使业主回想起其作为曾经的住宅的记忆。建筑的形式使人们想起了魔幻的山洞或者可弹跳的城堡。与原有的生产线式排列的住宅相比，新建的立体式木结构展现出一座具有更加有机的动态造型的幼儿园，立面和屋顶使用了常见的柔软的屋面材料，如沥青屋面板来进行防水，同时这种材料还将这座结构包裹起来，如同纸质的外壳一般。

为了增添一份趣味性，幼儿园内还设有一条滑梯，从二层向下通至建筑前方的室外游乐场和开放的庭院。

Clover House Kindergarten

MAD architects have completed their first project in Japan, the Clover House kindergarten. Located in the small town of Okazaki, the school's setting boasts views of the paddy fields and mountains, characteristic of the Aichi Prefecture.

The kindergarten was originally operated out of the old family home of siblings Kentaro and Tamaki Nara, which soon became too small and unfit for expanding their educational goals. The siblings desired to create a modern educational institution where children could feel as comfortable as they do in their own homes, allowing them to grow and learn in a nurturing setting.

MAD was commissioned by the family to transform their old two-story family house into a fully developed educational institution. The transformation started with an investigation of the existing 105 sqm house. Like the surrounding houses, this wooden building was first constructed as a standard prefabricated house. To keep the construction costs to a minimum,

©Rasmus Daniel Taun (courtesy of the architect)

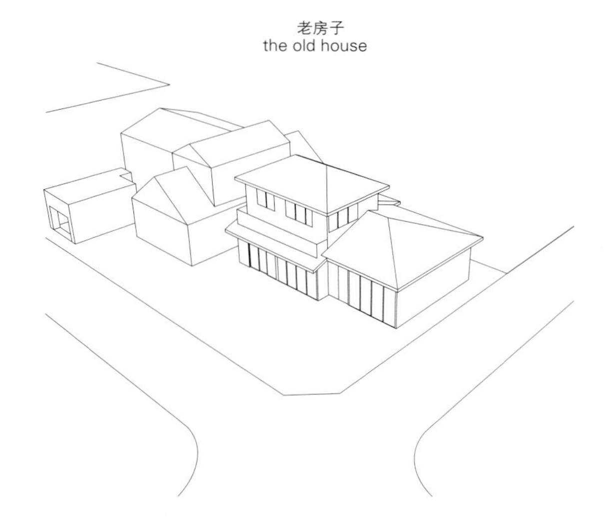
老房子
the old house

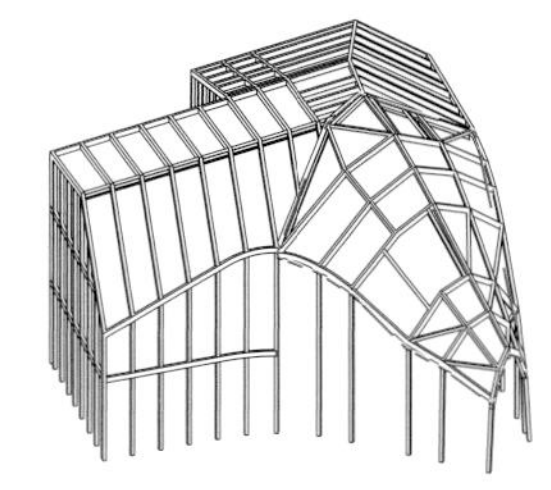

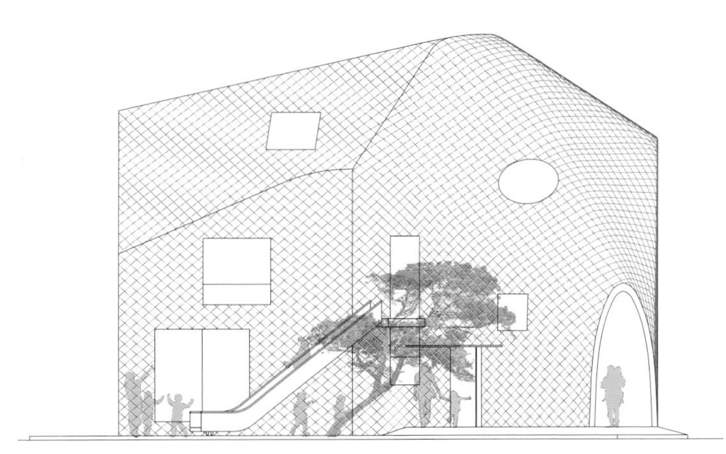
东南立面 south-east elevation

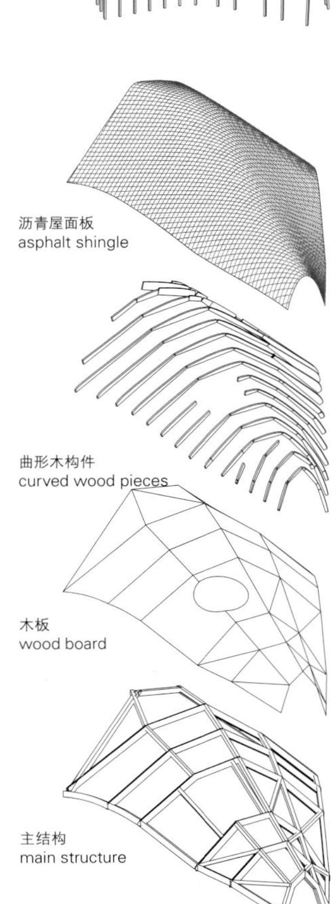

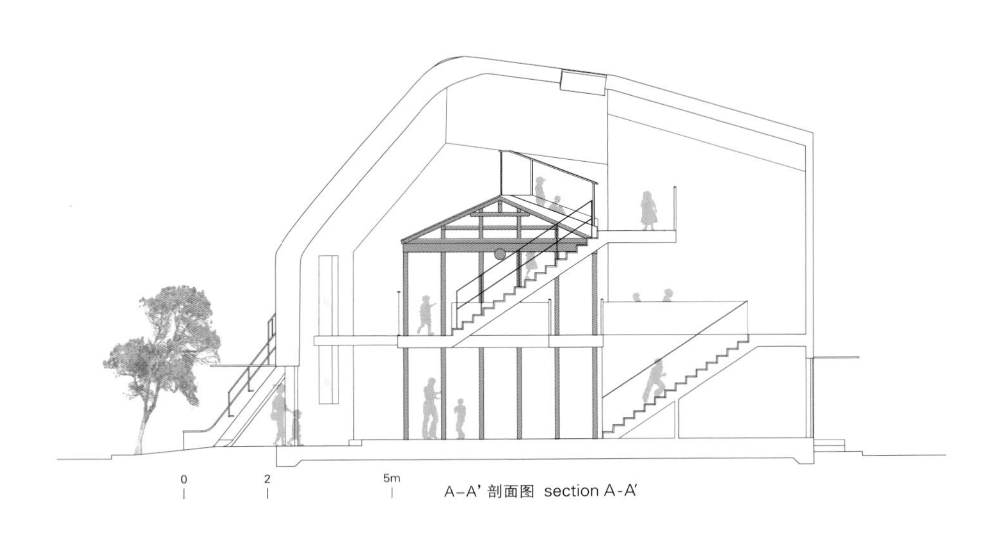

A–A' 剖面图 section A-A'

細井医院
ここ入る

项目名称：Clover House / 地点：Okazaki, Aichi, Japan / 建筑师：MAD architects / 首席建筑师：Ma Yansong, Yosuke Hayano, Dang Qun / 设计团队：Takahiro Yonezu, Yukan Yanagawa, Hiroki Fujino, Julian Sattler, Davide Signorato / 施工方：Kira Construction Inc. / 结构工程师：Takuo Nagai / 客户：Kentaro Nara, Tamaki Nara / 建筑类型：Kindergarten, Residence / 用地面积：283m² / 总建筑面积：134m² / 有效楼层面积：300m² / 设计时间：2012—2016 / 施工时间：2015—2016 / 摄影师：©Fuji Koji (courtesy of the architect) (except as noted)

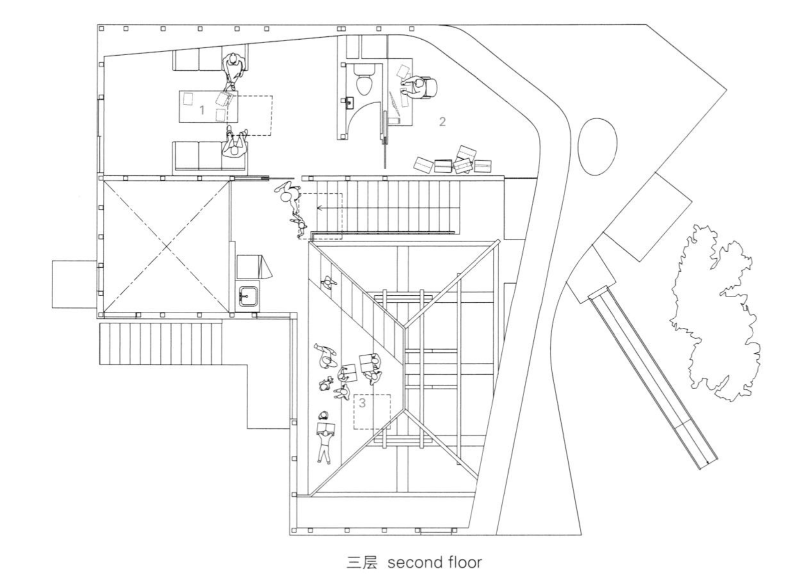

1. 起居室
2. 学习区
3. 阅读楼梯

1. living room
2. study
3. reading stairs

三层 second floor

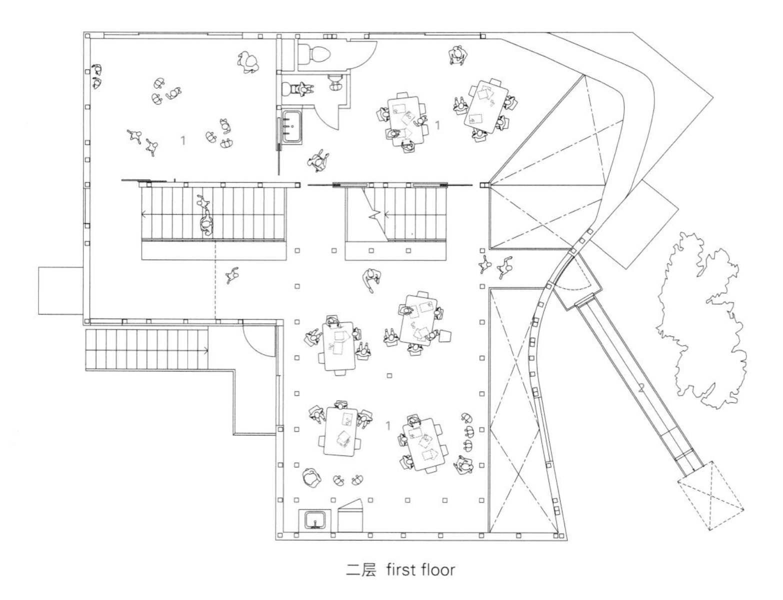

1.教室
2.滑梯

1. classroom
2. slide

二层 first floor

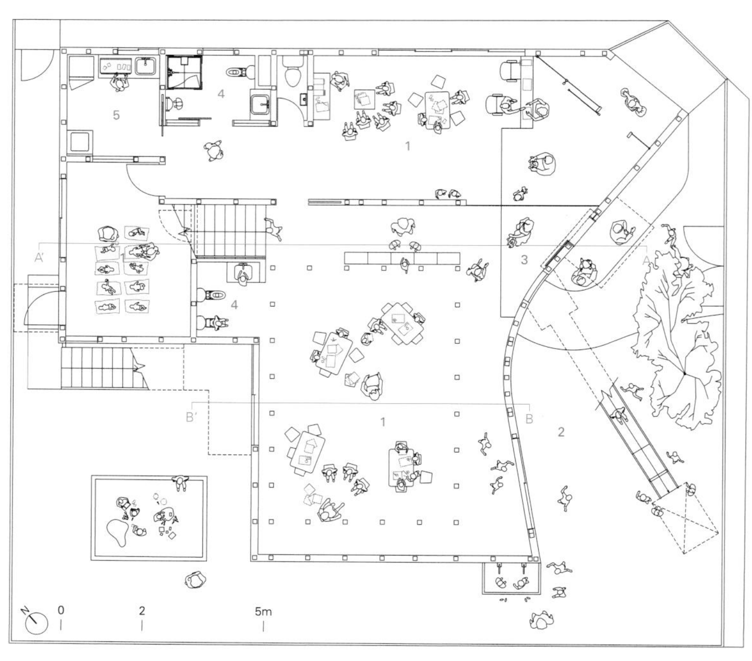

1.教室
2.游乐场
3.入口
4.浴室
5.厨房

1. classroom
2. playground
3. entrance
4. bathroom
5. kitchen

一层 ground floor

MAD decided to recycle the existing wood structure, incorporating it into the new building's design. The original wooden structure is present throughout the main learning area as a symbolic memory of Clover House's history. Its translucent and enclosed spaces easily adapt to different teaching activities. The windows, shaped in various geometries and recognizable to a child's eyes, allow sunlight to sift through and create ever-changing shadows that play with the students' curiosity and encourage imagination.

The new kindergarten's skin and structure wrap the old wooden structure like a piece of cloth covering the building's skeleton, creating a blurry space between the old and the new. The starting point of the Clover House Kindergarten is the signal pitched roof. This repurposed element creates dynamic interior spaces, and recalls the owners' memories of the building as their home. The form of the house brings to mind a magical cave or a pop-up fort. Compared to the original assembly-line residence, the new three-dimensional wooden structure presents a much more organic and dynamic form to host the kindergarten. The facade and roof utilize common soft roofing materials, such as asphalt shingles, to provide waterproofing, while wrapping up the whole structure in a sheath of paper-like pieces.

Adding to the sense of playfulness, there is a slide that descends from the first floor of the building to an outdoor play area and an open courtyard in front of the building.

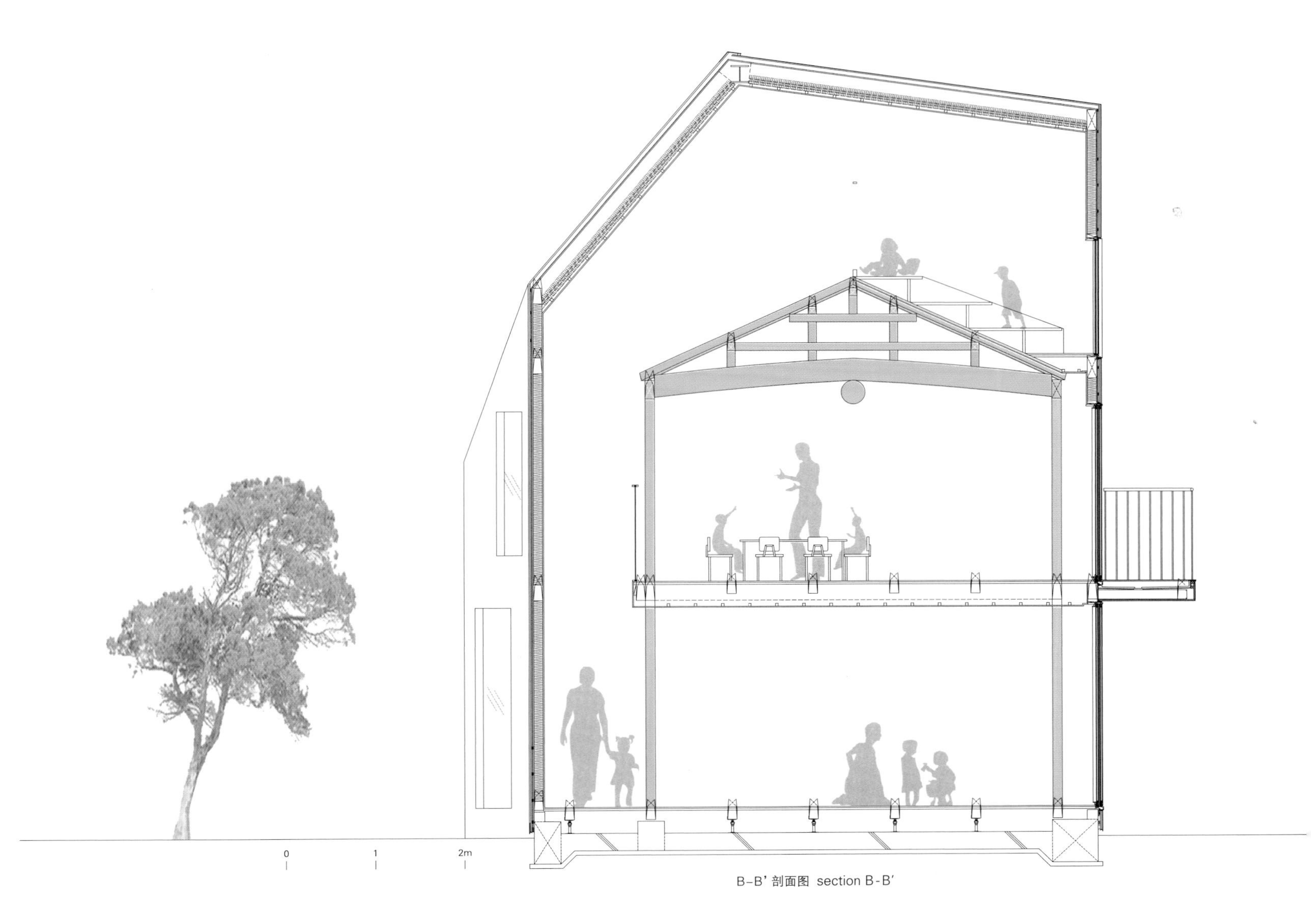

B-B' 剖面图 section B-B'

Jerry 住宅是业主 Sahawat 夫妇及四个孩子位于七岩海滩的度假住宅。

跑、爬、藏、悬挂、甚至下落等身体动作在日常生活中并不常见。Onion 事务所将 1940 年上映的美国动画短片"猫和老鼠"作为该项目室内设计的参照标准。如果要为老鼠 Jerry 修建房子的话，那么它很可能会是一座巨大的奶酪外形的房子，里面设有帮助自己躲避 Tom 猫追赶的"莫比乌斯圈"逃跑路线。Jerry 住宅被设计成一个深受孩子们喜爱的家园，同时使父母在七岩海滩的度假更有意义。

Onion 事务所的设计焦点是各通道的连接。除了已有的楼梯和斜坡之外，建筑师还在中央大厅内增加了垂直流线。该空间现被称为"主生活厅"。Onion 事务所用定制的绳网打造了五层拥有不同倾斜度、高度及尺寸的水平网层。五层绳网在三层高的生活厅内作为地板的延伸区，并且分别在每一层的不同位置设有一个孔洞。父母和孩子们可以通过金属楼梯从地面层爬到四层的一间儿童房。由于有下层绳网的保护，孩子们跳出床外或掉进孔洞都不会产生危险。父母和孩子可以抓住部分向下倾斜的绳网来玩耍。抬头看向天花板上的镜面时，吊挂在绳网上会产生悬浮在空中的感觉，层叠的网层看起来比它们实际的高度更高。

位于四层的卧室内另设其他流线。一条"隧道"将墙体内的两个内陷卧室连接起来。孩子们需要通过梯子或台阶进入卧室。他们还可以由第五层网上的隐藏滑动门从主生活厅直接到达卧室。若孩子们将房门从内部锁上，那么进入房间的唯一途径就是只能容纳孩子通行的"隧道"，大人难以进入。

颜色也是强化视觉体验的方式。

Onion 事务所将公共区漆成白色。而卧室则用不同的颜色打造了不一样的个性色彩。三层的主卧采用芥末黄色，客房为淡紫色。隧道和睡眠区域为蓝色。四层的男孩卧室为淡绿色。当孩子们在玩耍的时候，还可以通过房间的颜色来识别各自的房间。

Jerry住宅

Onion

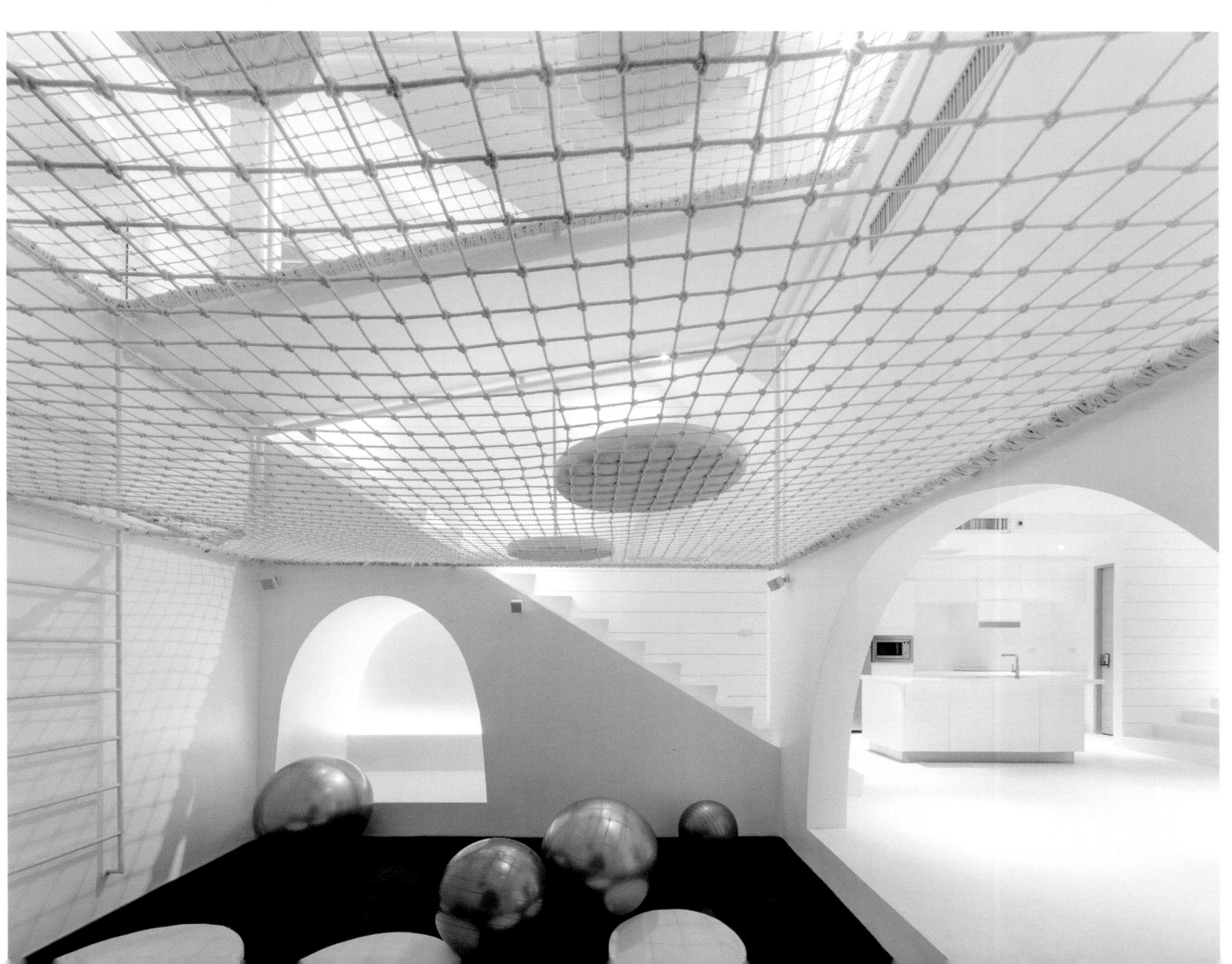

Jerry House

Jerry House is the holiday home at Cha Am Beach of Mr. and Mrs. Sahawat who have four sons.

Running, climbing, hiding, hanging and even falling are the proposed physical activities which exceed the boundary of common lifestyle. This is why the architects base this interior design project on an American animated series of short films created in 1940, "Tom and Jerry". If the house of Jerry Mouse was built, it might appear like a big piece of cheese in which the Möbius strip would be a desirable route for Jerry to run away from Tom Cat. Jerry House is thought of an object that children would yearn for and increases the meaning of having a holiday for the parents.

Central to the architects' concern is the linkage of passageways. They add an extra set of vertical circulation at the central hallway to the existing staircases and ramps. This space is now called the "main living hall". They use the customised

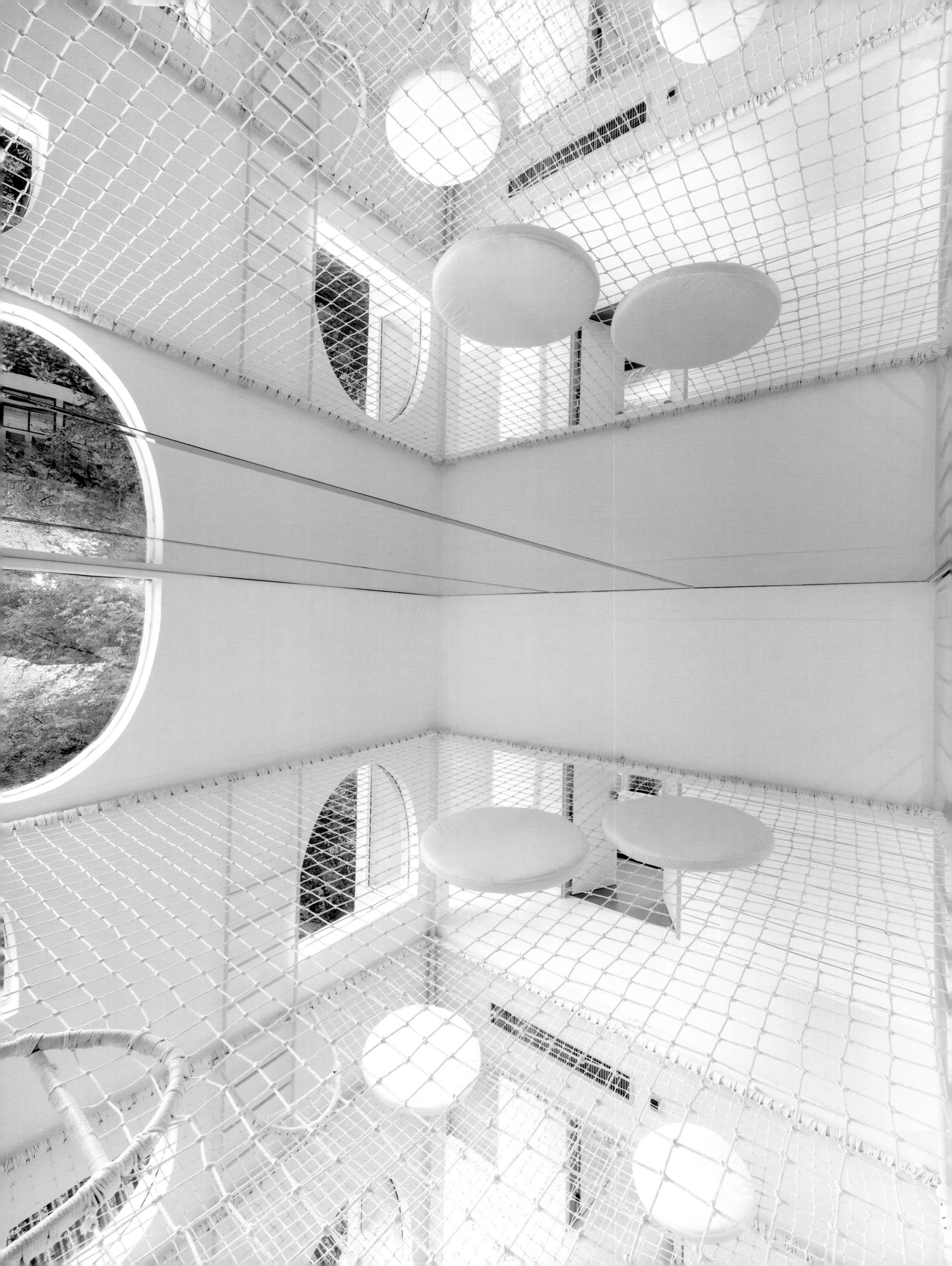

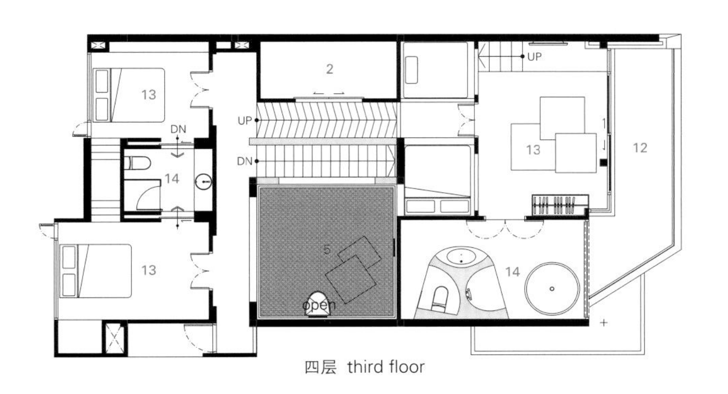

四层 third floor

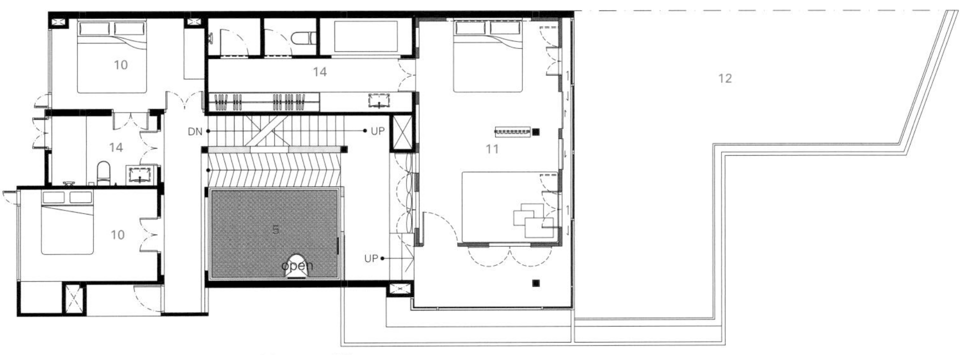

三层 second floor

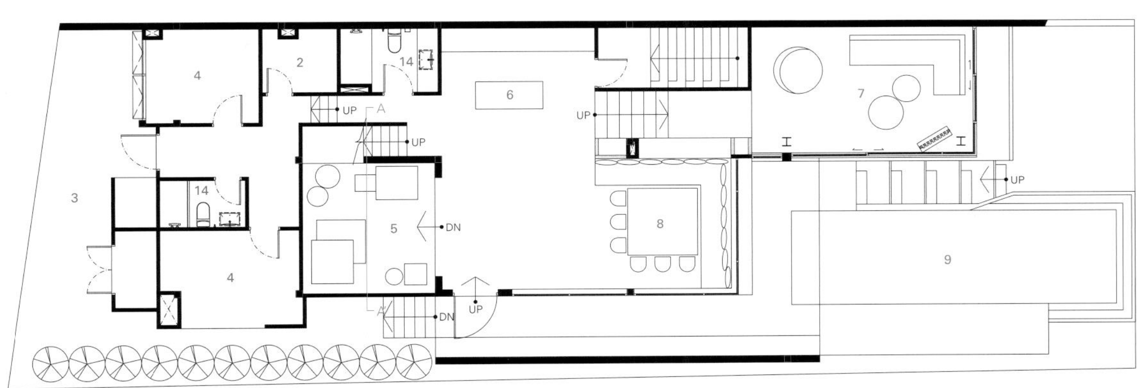

二层 first floor

1.停车场
2.储藏室
3.厨房&服务区
4.佣人房
5.主生活厅
6.备餐间
7.起居室
8.餐厅
9.泳池
10.客房
11.主卧室
12.平台
13.儿童卧室
14.浴室
15.更衣室

1. parking
2. storage
3. kitchen & service area
4. maid room
5. main living hall
6. pantry
7. living room
8. dining area
9. swimming pool
10. guest bedroom
11. master bedroom
12. deck
13. kid bedroom
14. bathroom
15. dressing room

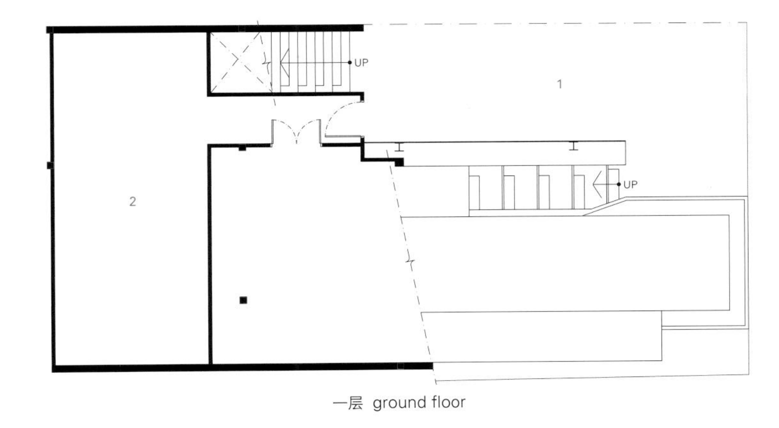

一层 ground floor

项目名称：Jerry House / 地点：Cha Am Beach, Thailand / 建筑师：Arisara Chaktranon & Siriyot Chaiamnuay_Onion
客户：Patta Sahawat / 总建筑面积：435m² / 竣工时间：2014 / 摄影师：©Wison Tungthunya (courtesy of the architect)

nets to build five layers of horizontal planes of different slopes, heights and sizes. They act almost as the extension of floors within this three-storey high living hall. Each layer of nets has a hole and each hole is set in different locations. Children and adults can climb up the metal ladders from the ground floor to one of the son's bedrooms on the third floor. Jumping out of bed and falling down the holes are safe because the players will be protected by the lower layers of nets. The players can grasp parts of the nets that are slanting down to fit their heights. When looking upward, the mirrored film attached to the ceiling gives the illusions that the players are floating in

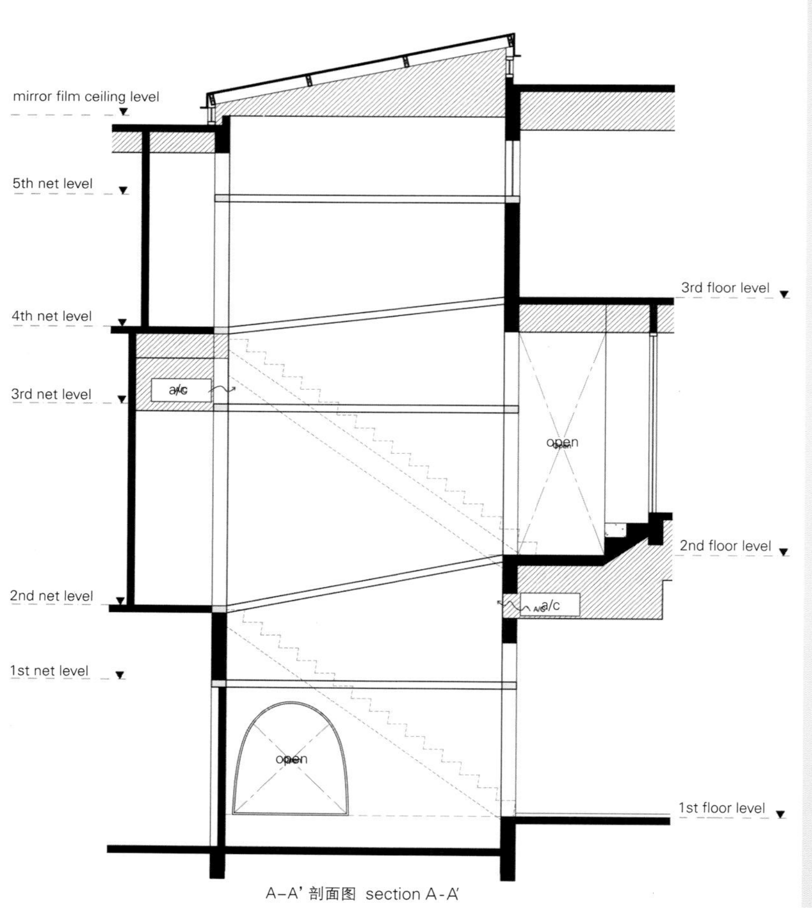

A–A' 剖面图 section A-A'

the air and the overlapping layers of net are higher than they actually are.

The bedrooms on the third floor have another type of circulation. It is the tunnel that links between two sunken spaces on the walls for sleeping in. In order to go to bed, the sons may climb the ladder or walk up the steps inside the bedroom. They may also go to bed directly from the main living hall through the hidden sliding door above the fifth level of the net. If the door of the chosen room is locked from the inside, the only access left will be the tunnel and that is difficult for the adults to get in.

Colors are the mean to increase the visual experiences. Onion paints the public areas in white. The bedrooms, on the other hand, are personalised by colors. Mustard yellow is for the master bedroom and light purple for the guests' bedroom on the second floor. Blue is for the interior of the tunnels and sleeping areas of two sons. Light green is for the boys' bedrooms on the third floor. When playing, the children tend to identify themselves with the room colors.

Hanazono幼儿园和托儿所

Hibinosekkei + Youji no Shiro

这座建筑位于距离日本东京 2000km 的宫古岛。宫古岛属于亚热带海洋气候区，被海洋和珊瑚礁所包围。由于地处台风盛行区的闷热且潮湿的环境，因此该建筑设计需要自由封闭和开放，来用于遮阳和通风。

由于台风会对施工结构产生一定的影响，因此该平面的规划将台风的因素考虑在内，同时还要求其能遮挡阳光。我们计划将一层设计为公共空间、工作室以及举办创意活动的画室，其中，画室对儿童发挥创意产生了重要的作用。

二层是一处更加亲密的私人区域，这里设有看护室和画册阅读室。建筑师利用一层狭窄的地势，将游乐场－工作室－画室－庭院－餐厅连接为一个整体，使之形成一处通风的大型空间。建筑使用了该地区内传统的红木瓦。然而，整体结构还是采用了钢筋混凝土，以防止台风的侵袭。幼儿园外围的低檐篷和围屏（由带孔洞的混凝土体块建成）采用的是传统建筑使用的混凝土材料，可以保护建筑免受飞来物品的碰撞以及阳光的照射，同时还保证了视野和通风。幼儿园外墙砖的颜色与传统的红木瓦相似，以融入该地区的建筑风格。

工作室与游乐场相连，而画室连接着庭院，餐厅则被露台包围。这里设有一座庭院，庭院的内外也是相连的，是一处非常舒适的活动空间。孩子们可以在工作室练习唱歌，在画室画画，在餐厅享用舒适的午餐的同时感受微风。

在游乐场和餐厅花园，孩子们还可根据当地植物结出的果实来感受四季的变化。工作室和画室内都可举办研讨会，孩子们在此聆听受邀的外部教师的讲解。这里，幼儿园负责人、父母和周边居民都可以在此一起工作，使这座幼儿园不仅仅扮演着教育孩子的角色，还是一处展现各种活动场景的生活区。

Hanazono Kindergarten and Nursery

This building is located in Miyakojima, approximately 2000 km away from Tokyo. Miyakojima belongs to a subtropical oceanic climate, surrounded by blue sea and coral reef. The building needed to be closed to make shade and to open for ventilation, due to the hot and humid climate in this typhoon-prone region.

The plan was made to accommodate typhoons and shut out the sunlight by analyzing how wind affects construction. We planned the first floor as public spaces, a studio and an atelier for creative activities which gives great importance to bringing up creativity in childhood.

On the second floor, a more intimate and private area, the childcare rooms and the area for picture books reading are located. Taking advantage of the narrow site on the first floor, the play ground - studio - atelier - courtyard - dining in a sequence are placed continuously and these become one large space which wind blows through when windows are opened. The traditional wooden red roofing tile seen in this area is used. However, the overall structure is steel reinforced concrete construction that endures a typhoon invasion. Around

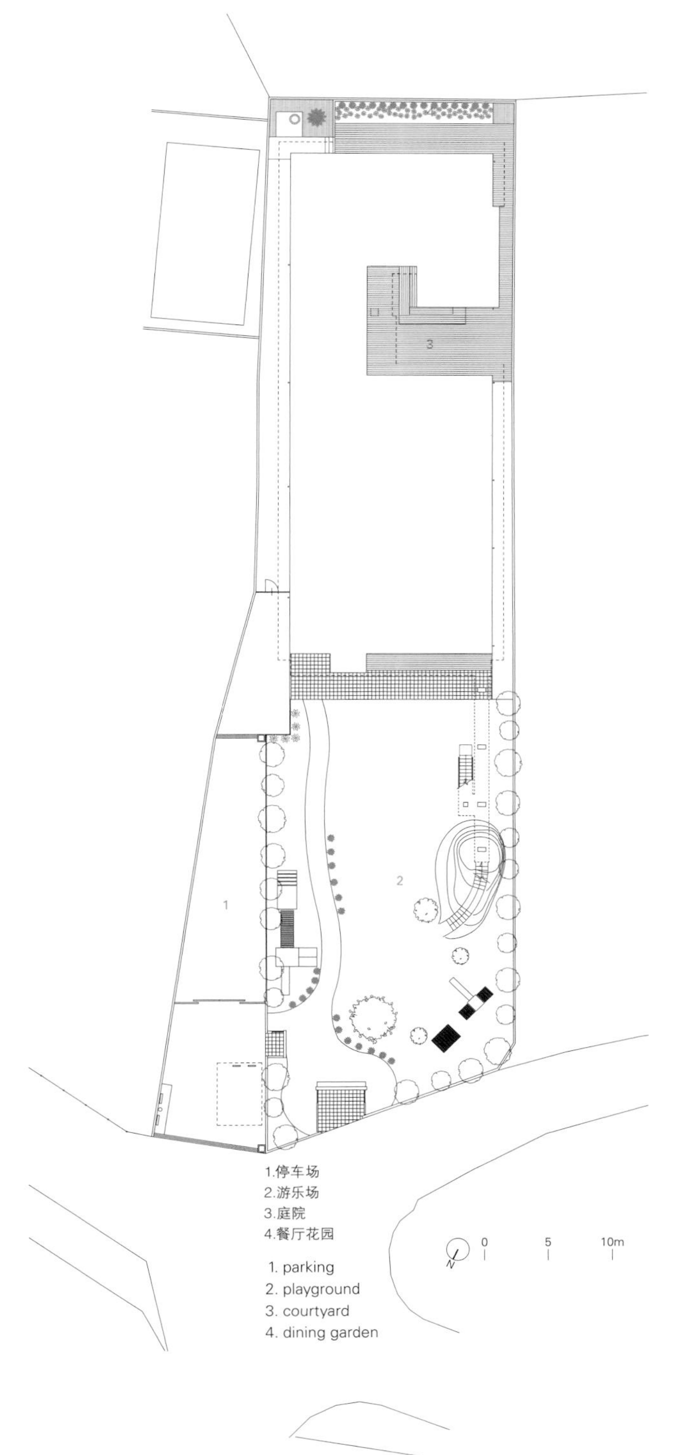

1.停车场
2.游乐场
3.庭院
4.餐厅花园

1. parking
2. playground
3. courtyard
4. dining garden

the outer perimeter of the building, low canopies and screens made by concrete blocks with holes, the original building material in this area, protect the building from flying objects and shut out the sunlight while maintaining view and wind. The color of the outer wall tile was chosen so that it would be similar to the traditional red roof tile color and blend into the region.

The studio is connected with a playground and the atelier with a courtyard; the dining is surrounded by a terrace. Another courtyard is a comfortable space where inside and outside are connected. Children play with sound in the studio and do productive activities in the atelier, and can enjoy lunchtime while feeling the cool breeze.

In the playground and dining garden, seasonal change can be felt by bearing fruits from the traditional plants. Workshops are held in the studio and atelier with lecturers invited from outside of the nursery. From this nursery, manager, parents and neighborhood work together and transmit various situations to the center beyond the limit of the child education.

项目名称：Hanazono Kindergarten and Nursery / 地点：Okinawa, Japan / 建筑师：HIBINOSEKKEI + Youji no Shiro / 用地面积：1,846.6m² / 总建筑面积：596.6m² / 有效楼层面积：1,107.6m² / 设计时间：2013.10~2014.2 / 施工时间：2014.5~2015.3 / 摄影师：©Ryuji Inoue-Studio Bauhaus (courtesy of the architect)

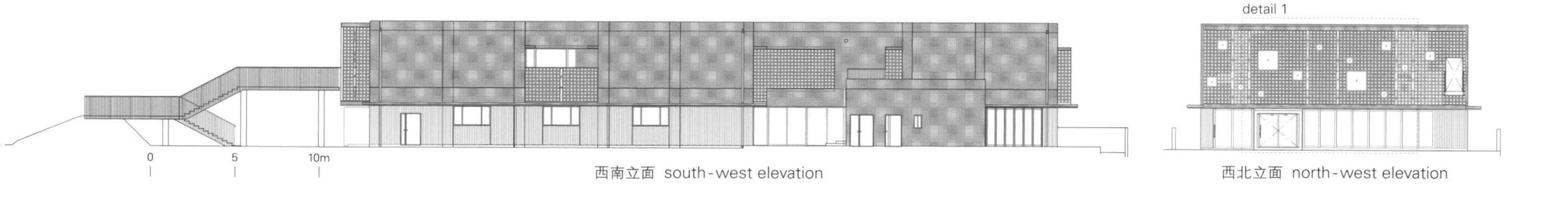

西南立面 south-west elevation

西北立面 north-west elevation

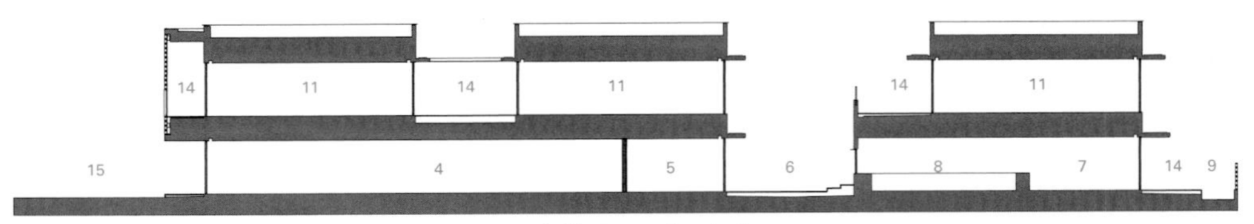

A-A' 剖面图 section A-A'

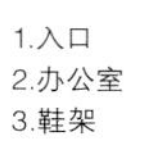

1.入口
2.办公室
3.鞋架
4.工作室
5.画室
6.庭院
7.餐厅
8.厨房
9.餐厅花园
10.卫生间
11.看护室
12.画册阅读室
13.小型娱乐室
14.露台
15.游乐场

1. entrance
2. office
3. shoe shelves
4. studio
5. atelier
6. courtyard
7. dining
8. kitchen
9. dining garden
10. toilet
11. childcare room
12. picture book room
13. small playroom
14. terrace
15. playground

B-B' 剖面图 section B-B'

C-C' 剖面图 section C-C'

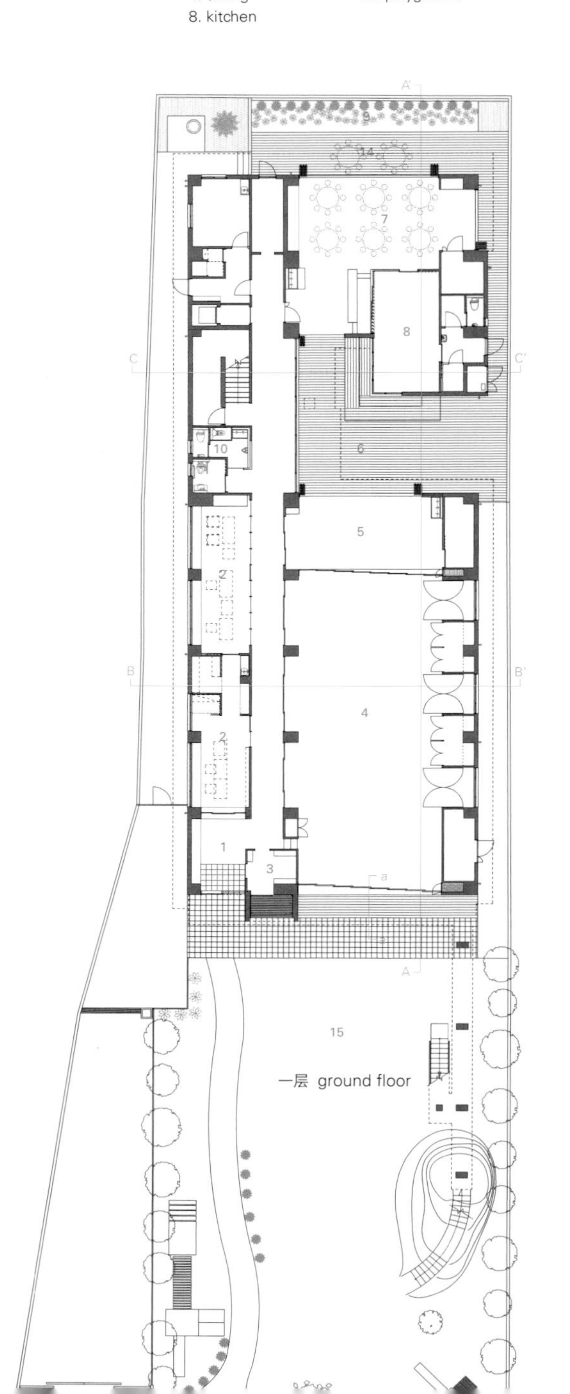

一层 ground floor

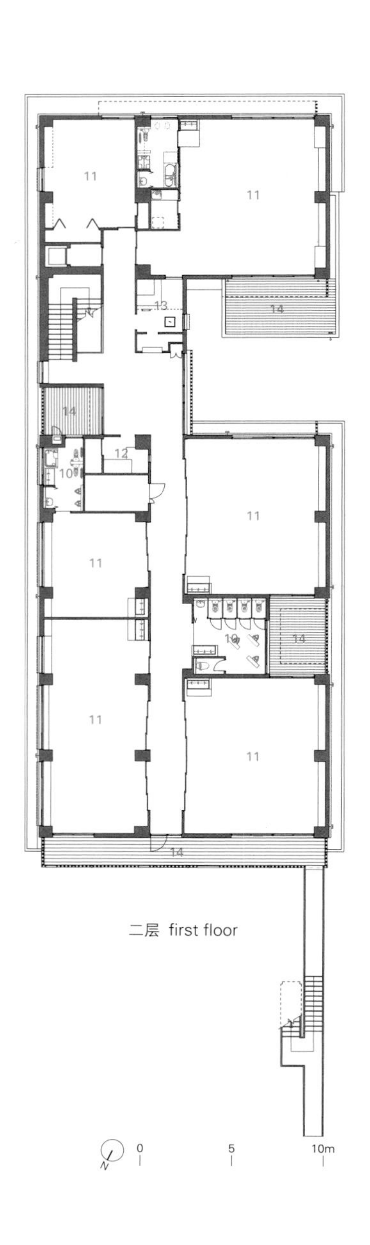

二层 first floor

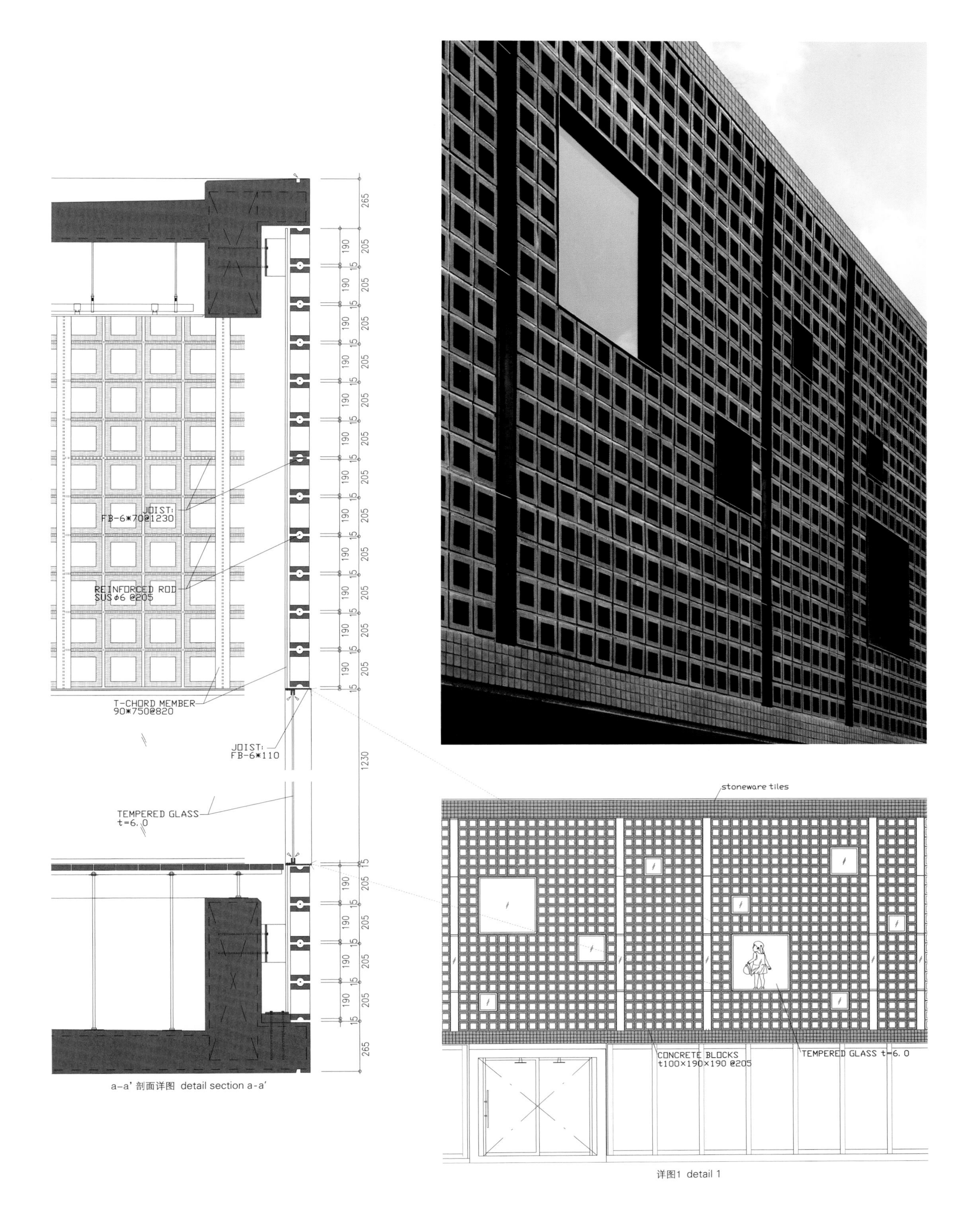

a-a' 剖面详图 detail section a-a'

详图1 detail 1

耶路撒冷以色列博物馆的青年艺术教育中心

Ifat Finkelman + Deborah Pinto Fdeda

这座位于耶路撒冷以色列博物馆的青年艺术教育中心设有一个入口庭院，是游客（成人、孩子和团体）的主要聚会场所。庭院的周围都是林立的现代化混凝土－石材建筑。而本项目是对现有庭院的改造，将游客自行活动的功能与明确规定的环境结合起来。

原有的松树是该项目的焦点。建筑师设计了一个小型带屋顶的结构，使孩子们拥有童年在树屋中一起玩耍的回忆。这里是孩子们捉迷藏和俯瞰风景的地方，位于倾斜的树干之上，从布局一丝不苟的博物馆周围环境中脱颖而出。这座树屋除了作为一个标志性存在物之外，还是其所属的这个连续的折叠结构的顶点。

2cm 厚的的 ipea 板材固定在轻质钢骨架之中，这种技术从上到下都营造了一系列的透明度。当该结构的表面逐渐向地面延伸时，便会转化为游乐场的地面。座位区作为一处独特的地形，覆盖了柔软的 EPDM 橡胶表面。所有这些设计都将位于地下的基础设施以及蔓延的大树根部系统隐藏起来。

在夜晚，这座树屋是唯一一处被点亮的区域，看起来好似漂浮在庭院的入口处。

The Youth Wing for Art Education, Israel Museum Jerusalem

The entrance courtyard of the Youth Wing for Art Education at the Israel Museum in Jerusalem is a main gathering place for visitors – adults, children and groups – within its modern concrete-and-stone architectural surroundings. The project – a renewal of the courtyard – combines a program open to interpretation by its users with a clearly defined context.

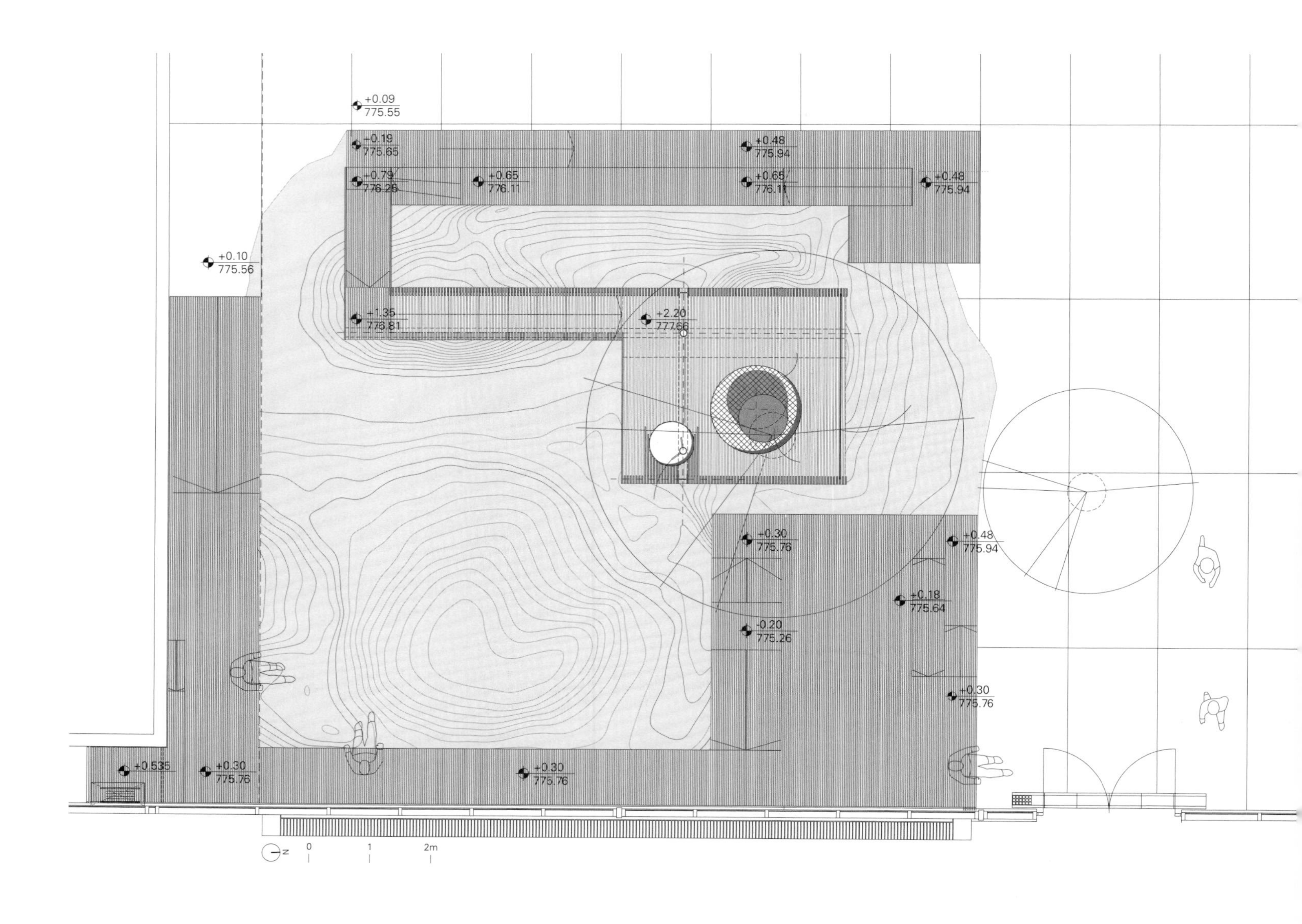

+0.09
775.55
+0.19
775.65
+0.48
775.94
+0.79
776.25
+0.65
776.11
+0.65
776.11
+0.48
775.94
+0.10
775.56
+1.35
776.81
+2.20
777.66
+0.30
775.76
+0.48
775.94
+0.18
775.64
-0.20
775.26
+0.30
775.76
+0.535
+0.30
775.76
+0.30
775.76
N
0
1
2m

The existing pine tree is the focus of the project. As a tribute to the childhood collective memory of a tree house, a small roofed structure where children can hide and over look out, is positioned high up the tilted trunk, raised above the meticulous surroundings of the museum. Along with its strong iconic appearance, it also functions as the peak of one continuous structural folded element.

The structural technique – 2 cm Ipea boards fixed to a light steel skeleton – creates a range of transparencies from top to bottom. While gradually transforming towards the ground, the element's surface becomes a playground; sitting elements frame a topography covered with a soft EPDM rubber surface. All these carefully hide the underground infrastructure as well as a widespread root system.

At night the house is the only element illuminated, and emerges floating above the courtyard's entrance.

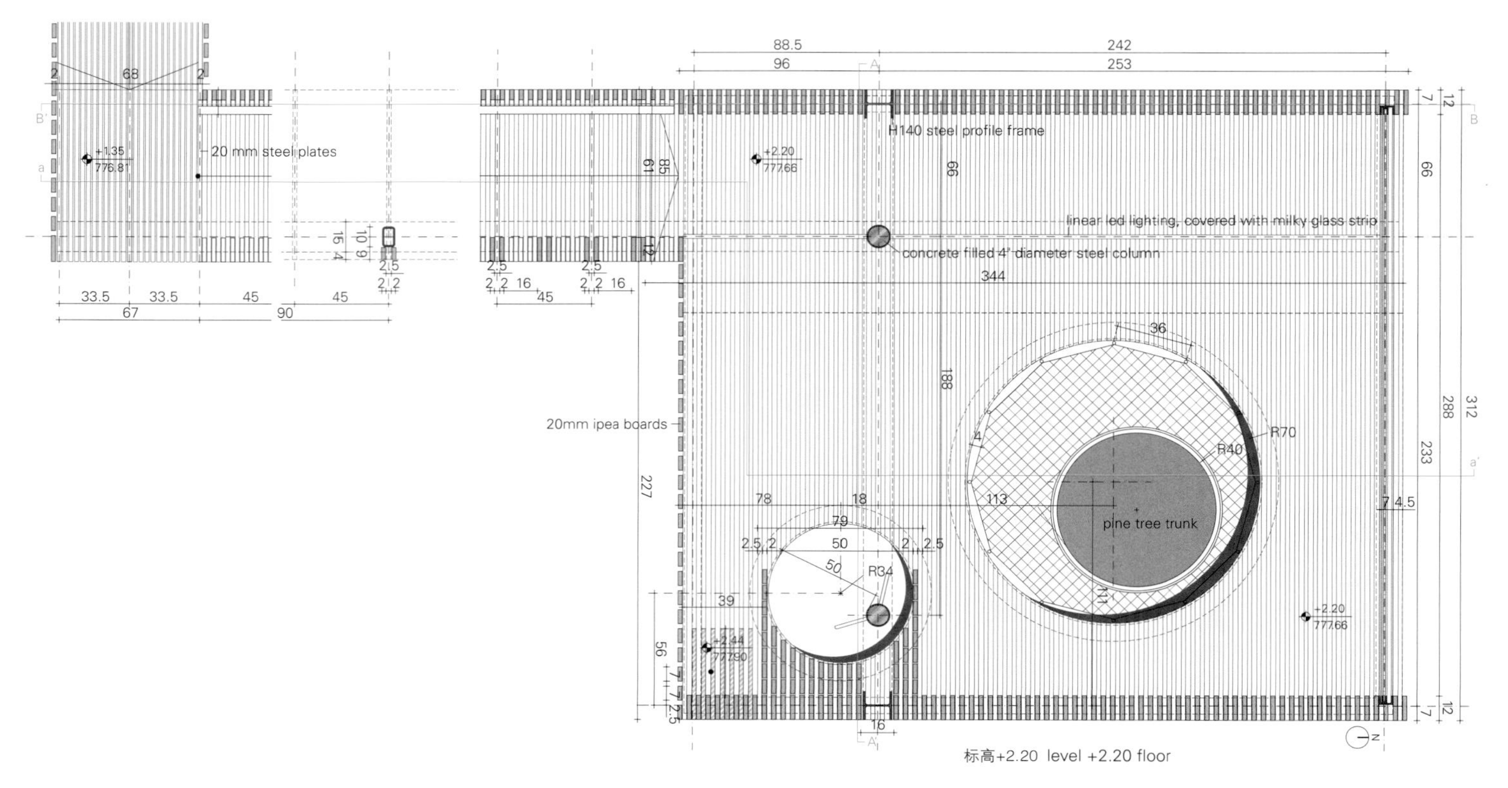
20 mm steel plates
H140 steel profile frame
linear led lighting, covered with milky glass strip
concrete filled 4" diameter steel column
20mm ipea boards
pine tree trunk
标高+2.20 level +2.20 floor

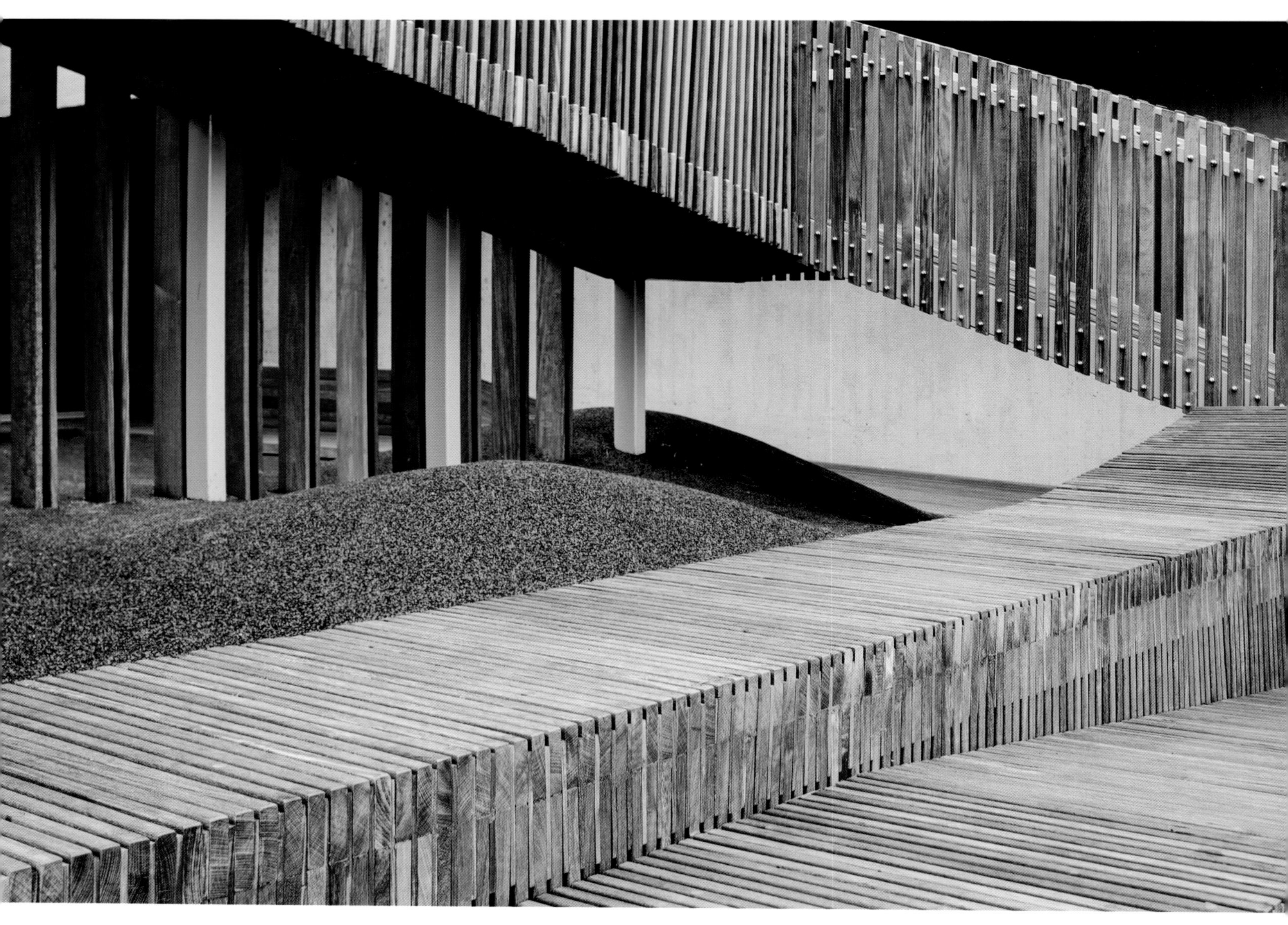

项目名称：The Youth Wing for Art Education, Israel Museum Jerusalem
地点：Derech Ruppin, Jerusalem, Jerusalem
建筑师：Ifat Finkelman, Deborah Pinto Fdeda
出资人：Lier Foundation
结构工程师：Ilan Ben David Engineers ltd.
安保：RSGD ltd.
通道设计：Inclusion ltd.
承包商：Greensky ltd.
EPDM地面设计：Politan Sport ltd.
钢结构设计：Bazelet ltd.
安全网设计：Pealton ltd.
照明工程师：Hila Meier, RTLD Lighting Design
客户：The Ruth Youth Wing for Art Education, the Israel Museum, Jerusalem
总建筑面积：150m²
竣工时间：2014
摄影师：©Amit Geron (courtesy of the architect)

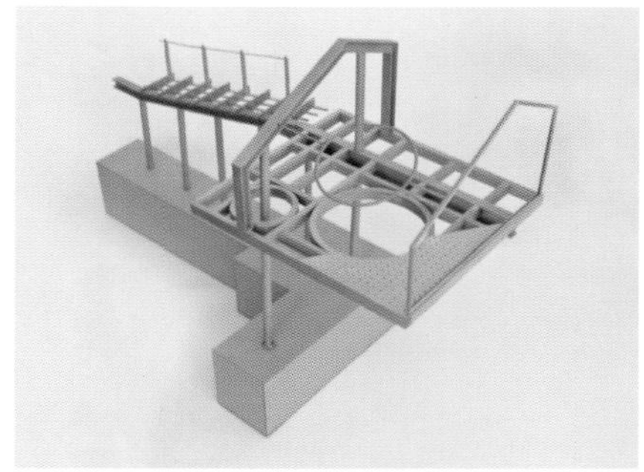

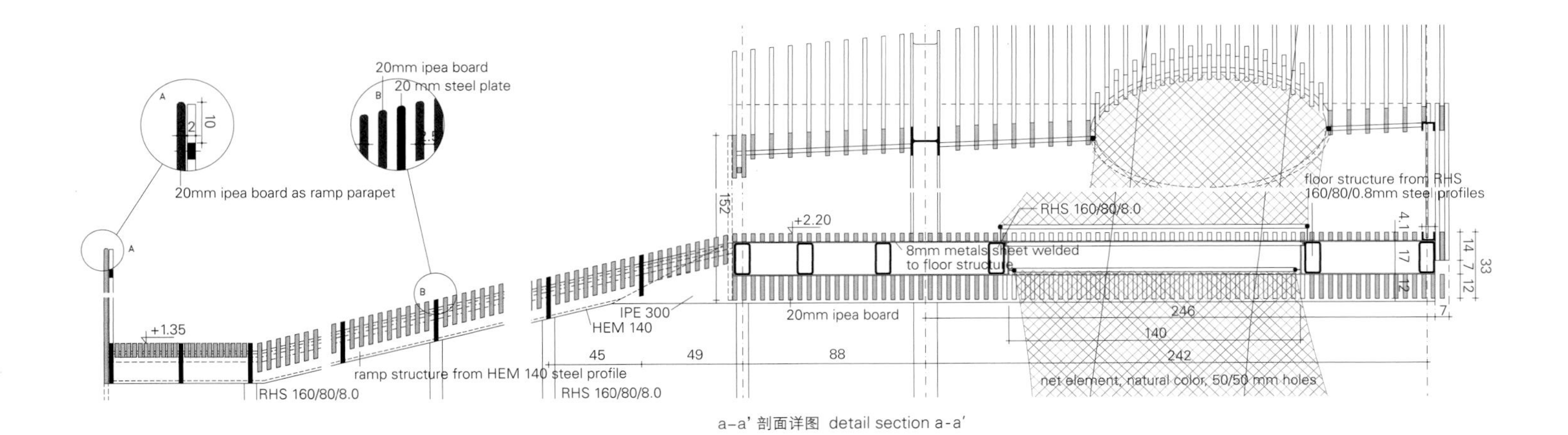

a-a' 剖面详图 detail section a-a'

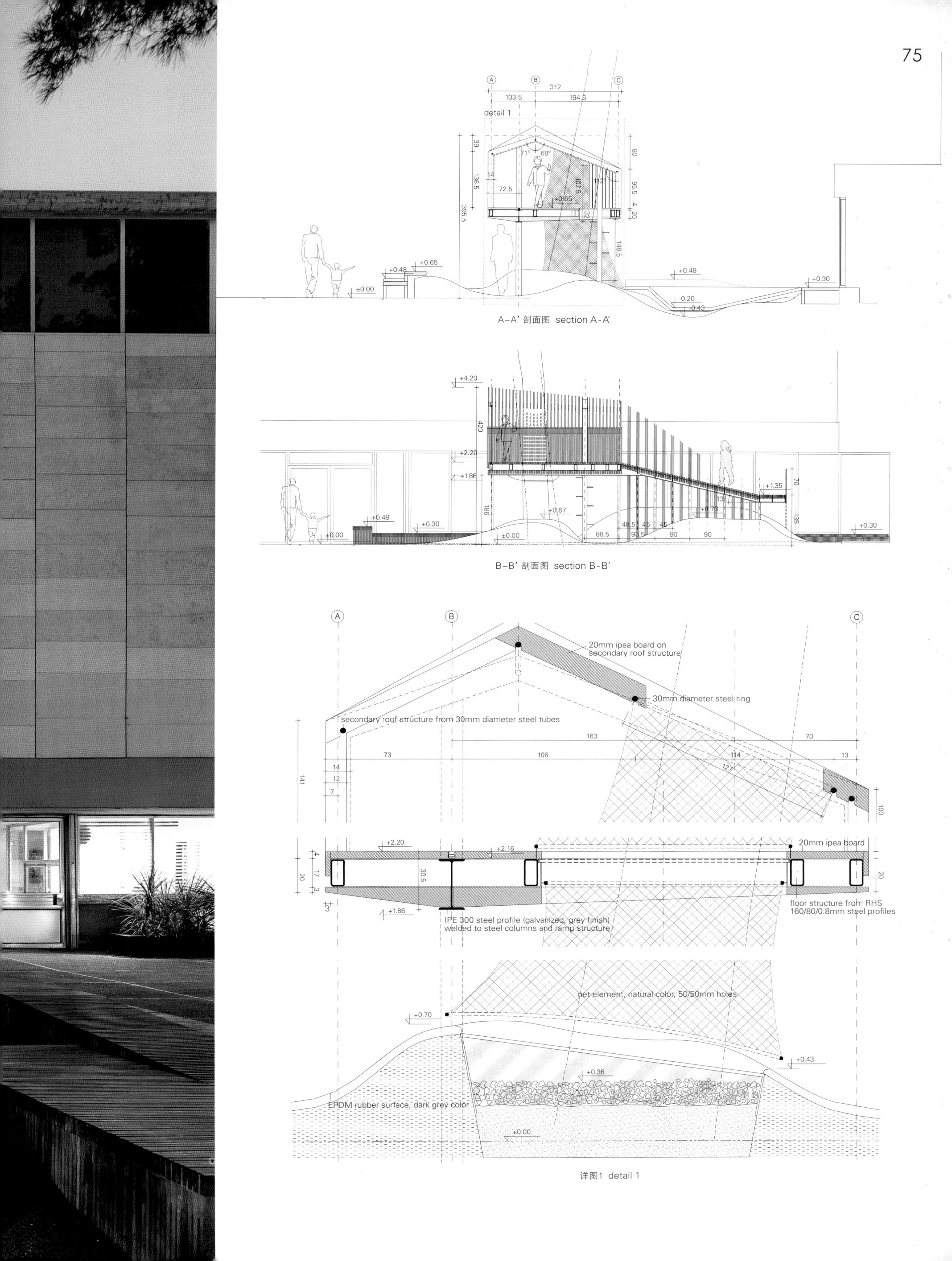

A–A' 剖面图 section A-A'

B–B' 剖面图 section B-B'

详图1 detail 1

Cassarate幼儿园

Bruno Fioretti Marquez

这座幼儿园位于一处名为 Cassarate 的街区内，该街区是在 19 世纪城市扩建时形成的。在过去的几十年里，Cassarate 街区得到了进一步的开发。这里最常见的建筑类型包括三至八层的住宅建筑，这些住宅的一层却作为商用。一些建筑与街道对齐，其他建筑则做后退设置，以形成一个前花园。这种城市布局在 19 世纪紧凑的城市风范和绿色景观区内的现代化乌托邦独立建筑风格之间摇摆不定，从而形成缺乏特色的城市空间。因此该设计的一个重要目标是建造一处明确规定的室外空间。

该建筑在城市街区的位置使其成为凸显城市特色的新结构。新建体量包含三座建筑：小学、体育馆(将和一座新大厅一起竣工)和幼儿园，这三座建筑将在两条街道之间交替与街道对齐。

由此在建筑和街道之间的开放空间形成了体育馆前的体育场、新幼儿园的游乐场以及小学的入口广场。

这三处空间都是通过人行道相连的，以展示校园的氛围。每处空间都设有一条不同的公共通道，其所使用的材料以及设计的绿化区和大众使用的家具都赋予自身一个鲜明的特点。

建筑的外表面在周围沥青表面环境的映衬下，突出了这些开放空间的特色。开放空间不仅仅供学生使用，同时也在放学之后，供公众使用，为邻里居民提供一处急需的会面场所。幼儿园的内部设有私人花园，每座花园内都栽种了独特的水果树品种。

学校还计划沿着其周边栽种其他的树木，来进一步凸显该区域现有的绿化区，同时将街区与 Cassarate 河在视觉上连接起来。

实际上，幼儿园建筑可被理解为一座单层的微型城市，在这里，每个教学团体都分配在带有私人花园和特殊的水果树的单独房子内。

这座幼儿园综合建筑犹如大型积木，共包含 56 个梯形模块，其中 35 个模块是儿童使用的供暖空间，13 个模块作为露天花园，最后 8 个模块的屋顶则是封闭的，作为中央走廊使用。

每个教学团体都分配了 5 个模块，分别用作一间衣帽间、一处卫生服务区、一间餐厅以及两间静室。其中两个模块可结合在一起，作为游戏区。此外，两个教学单元也经常共享这处结合区。

该综合体的模块结构为每个教学单元提供了 5 个相同的模块，且这些模块可随时结合起来。通过不同的模块结合，每个教学单元都被赋予了独特的特征，同时空间还具有了多样性。基础模块所形成的不规则几何外形所营造的复杂性同多样化排列性共同赋予了建筑强烈的可塑性和特征。

这些梯形基础模块包括预制的重型承重墙，墙体的外层为木板。屋顶也是由预制的木椽和支撑的木嵌板组装在一起的。计算机辅助的高水平预制技术可使非常复杂的立体屋顶表面精确地组合在一起。

模块化还使两段的施工期对现有幼儿园的使用造成最小的影响。室内装置和家具都计划使用木材来建造，以和承重结构更好地结合或整合在一起，使材料给人统一且和谐的印象。

Cassarate Kindergarten

The kindergarten is located in the neighborhood named Cassarate, a 19th century city expansion, which has been further consolidated in the last decades. The most common architectural typology consists of residence buildings between three and eight floors with commercial use on the ground floor. Some buildings place themselves on the street's alignment; others retire backwards to form a frontal garden. This pattern, oscillating between the 19th century compact city and the modern utopia of isolated buildings in the green landscape, generates an urban space lacking of specific character and quality. An unambiguous definition of the outer space was therefore an important goal of the building's design.
Its position on the city block creates a new structure that reinforces its urban quality. The new ensemble consists of three buildings blocks: primary school, gymnasium (to be completed with a new hall) and kindergarten, which alternate their street alignment between the two streets.
The resulting open spaces between the buildings and the streets provide a sport's field for the gymnasium, a playfield for the new kindergarten and the entrance's square of the primary school.
All three spaces are connected through pedestrian walks suggesting a "campus" atmosphere. Each space has a different public accessibility and also a specific identity defined by its materiality, green areas and public furniture.
The outer surfaces reinforce the specific characters of the open spaces in the background of the typical asphalted surface of the context. The open spaces are not only used exclusively by the pupils but also, after the school opening hours, open to public use providing a most needed meeting point for the neighborhood. The interior of the kindergarten complex allows space for intimate gardens each with its specific fruit trees.
Other trees are also planned along the perimeter of the

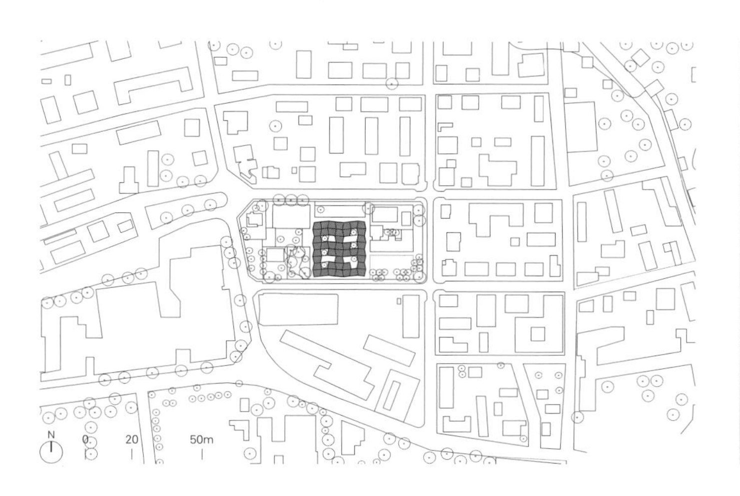

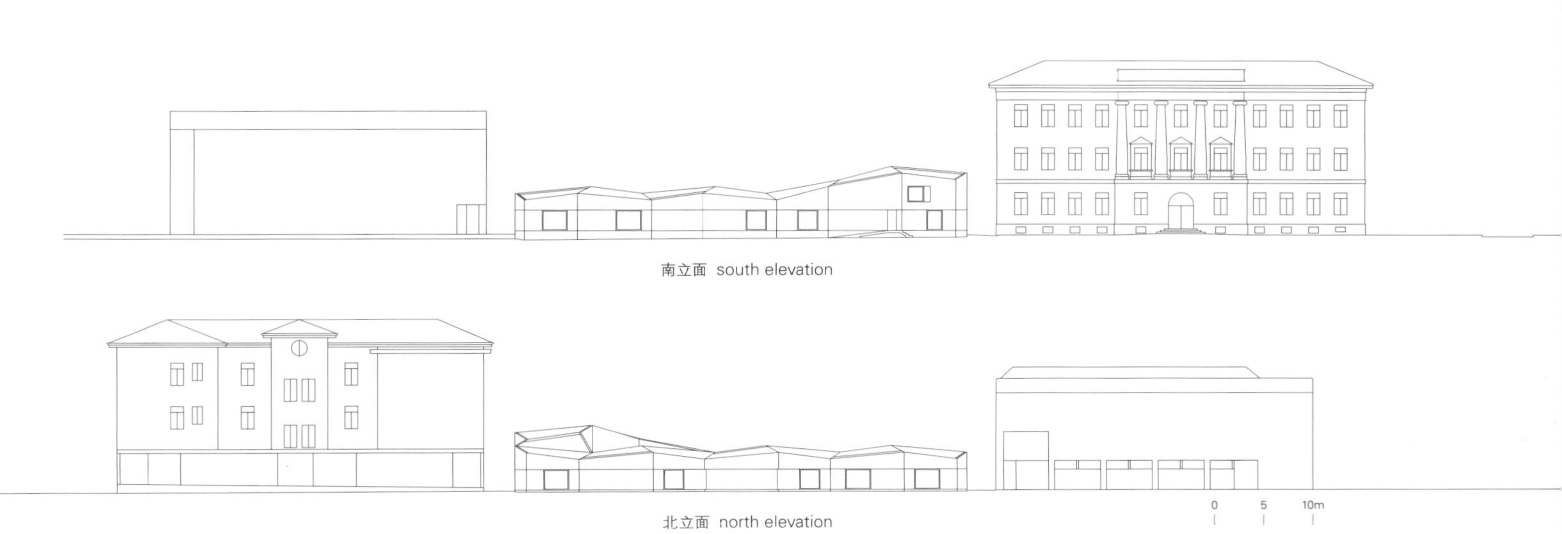

南立面 south elevation

北立面 north elevation

项目名称：Kindergarden Cassarate / 地点：Via Concordia 7, 6900 Lugano
建筑师：Bruno Fioretti Marquez Architekten
合伙人：Piero Bruno
项目负责人：Inken Blum, Regina Maria Münstermann, Fabian Wichers
出资人：Sidney Bollag, Vito Priolo
结构工程师：Zanini & Borlini SA
建筑物理：physARCH Sagl
建筑设备：Comunita di lavoro, A. Reichlin M.Gavazzini
电气工程师：C&C Electric SA
景观建筑师：Capatti Staubach
施工管理：Rolando Spadea Sagl
客户：Città di Lugano / Dicastero Edilizia Pubblica e Genio Civile
有效楼层面积：2.290m² (Kindergarten only)
体量：6,966m³ (Kindergarten only)
能源标准：Minergie P / 供暖：Thermal heat pump
造价：12,250,000 CHF(Incl. kindergarten, exterior, roofed schoolyard, renovation of school, professional fees)
竞标时间：2007 / 施工时间：2010—2014
摄影师：©Alessandra Chemollo / ORCH (courtesy of the architect)

school, to further reinforce the existing vegetation of the area, while contributing to a clear definition of the connection between the neighborhood and the Cassarate river.
Essentially the building for the kindergarten can be understood as a single story miniature city where each teaching-group is assigned to a house with its own garden and a particular fruit-tree.
The kindergarten complex, resembling a giant toy-block, consists of 56 trapezoidal modules, from which 35 provide heated spaces for the children, 13 open to the sky for the gardens, and 8 are roofed for the central distribution gallery.
Each teaching group consists of a combination of 5 modules:

1 for the wardrobe, 1 for the sanitary services, 1 as eating room, and 2 for quiet activities. 2 modules can be combined to provide spaces for game activities and are commonly shared by two teaching-units.
The modular structure of the complex provides every teaching-unit with exactly 5 same modules while allowing a new combination each time. Through the different combinations every group acquires a unique identity and at the same time contributes to the general diversity of the spaces. The complexity generated by the irregular spatial geometry of the basic modules and their many permutations provide the building with a strong plasticity and identity.

The basic trapezoidal module consists of prefabricated massive load bearing walls with an external cladding in wood. The roof is also assembled with prefabricated rafters and wood panels for bracing. The high level and computer-aided prefabrication allowed an accurate assemblage even of the very complex three-dimensional geometry of the roof's surface. The modularity also permitted a construction schedule in two phases with minimal interference with the existing kindergarten function. Interior fixtures and furniture were planned in wood allowing people to combine/integrate more with the load bearing structure, creating a single coherent material harmony.

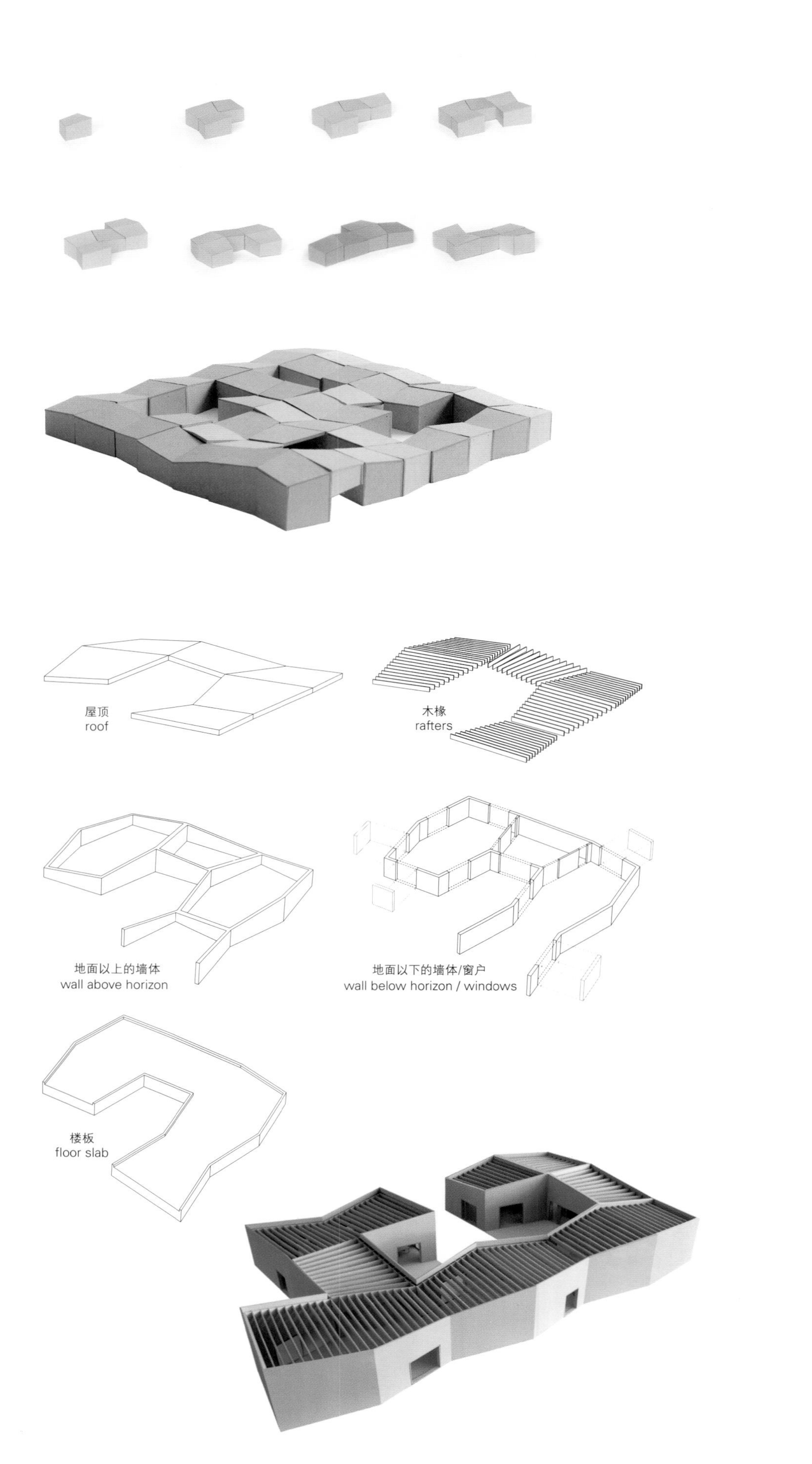

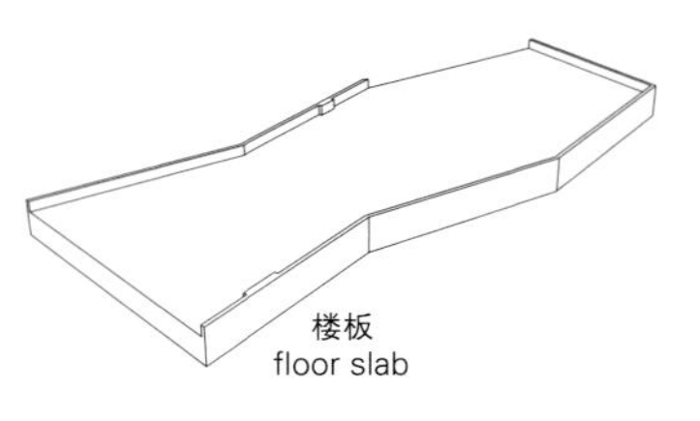
楼板
floor slab

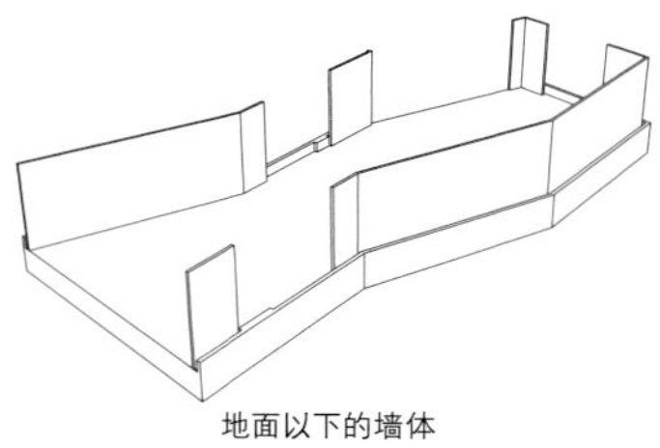
地面以下的墙体
wall below horizon

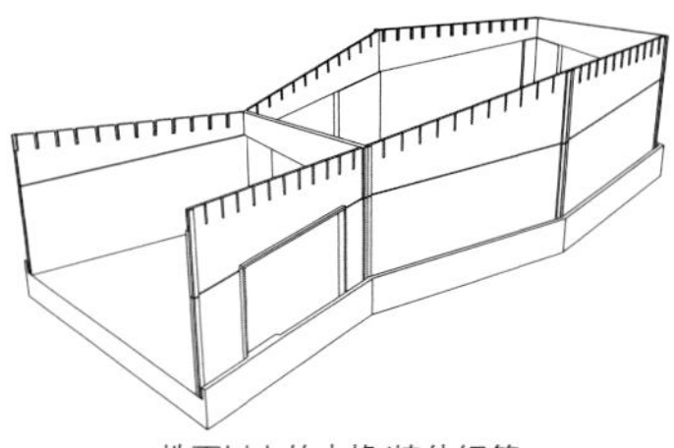
地面以上的木椽/墙体钢筋
rafters above horizon / reinforcement of walls

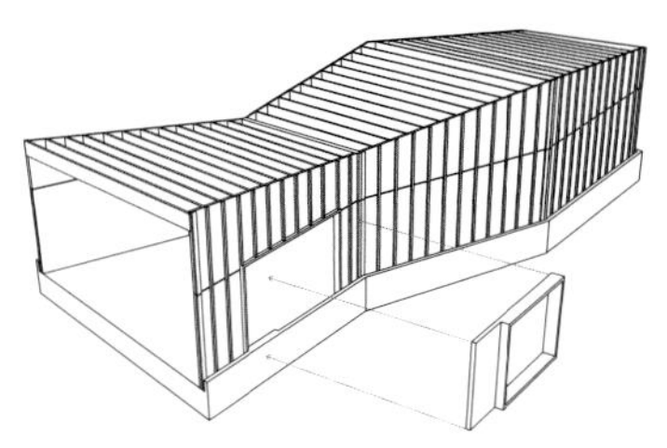
木椽/基础结构/窗户
rafters / substructure / windows

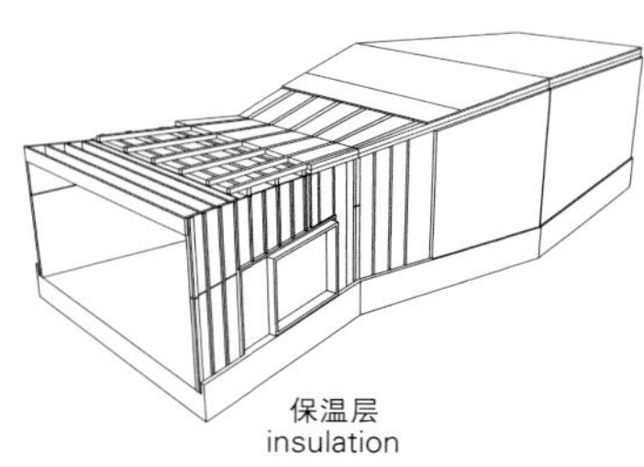
保温层
insulation

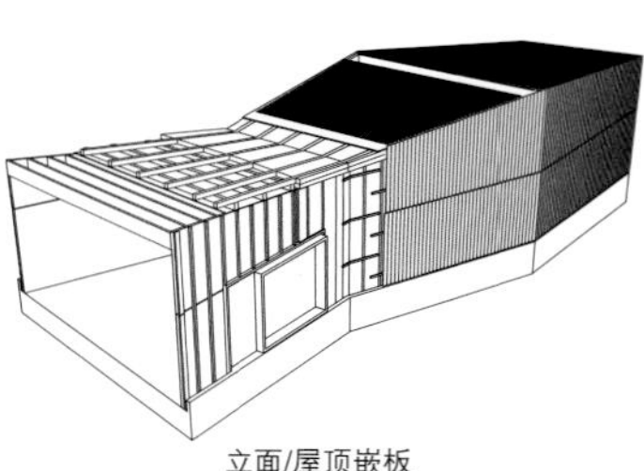
立面/屋顶嵌板
facade / roof panels

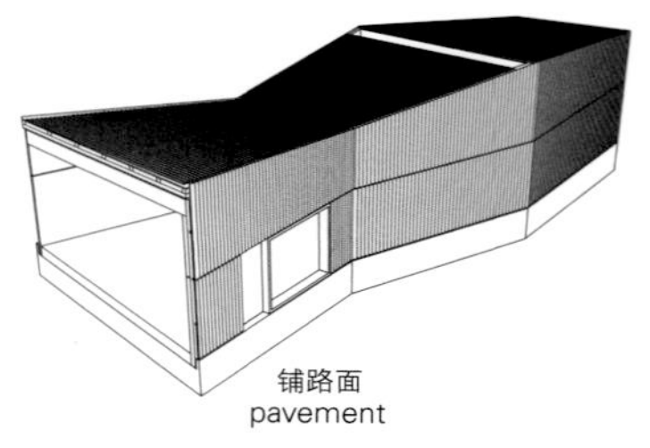
铺路面
pavement

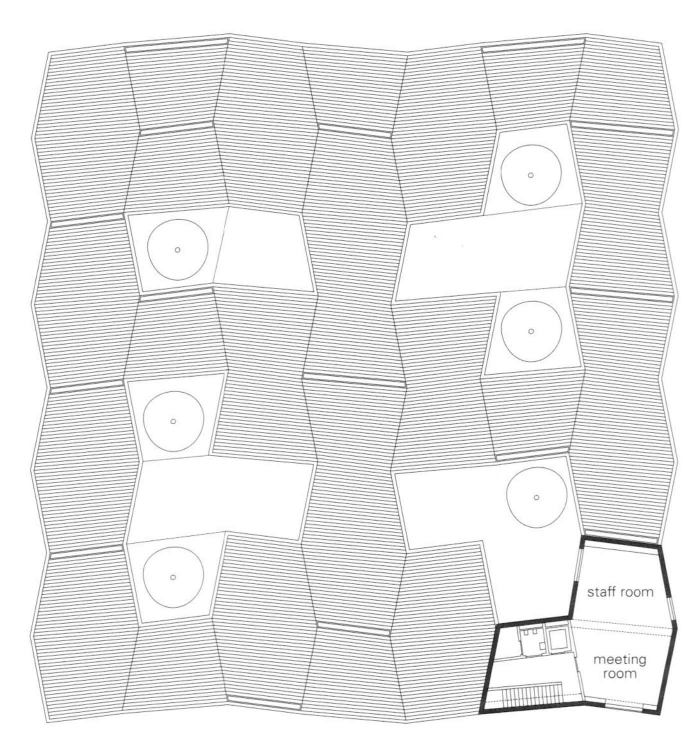

屋顶 roof

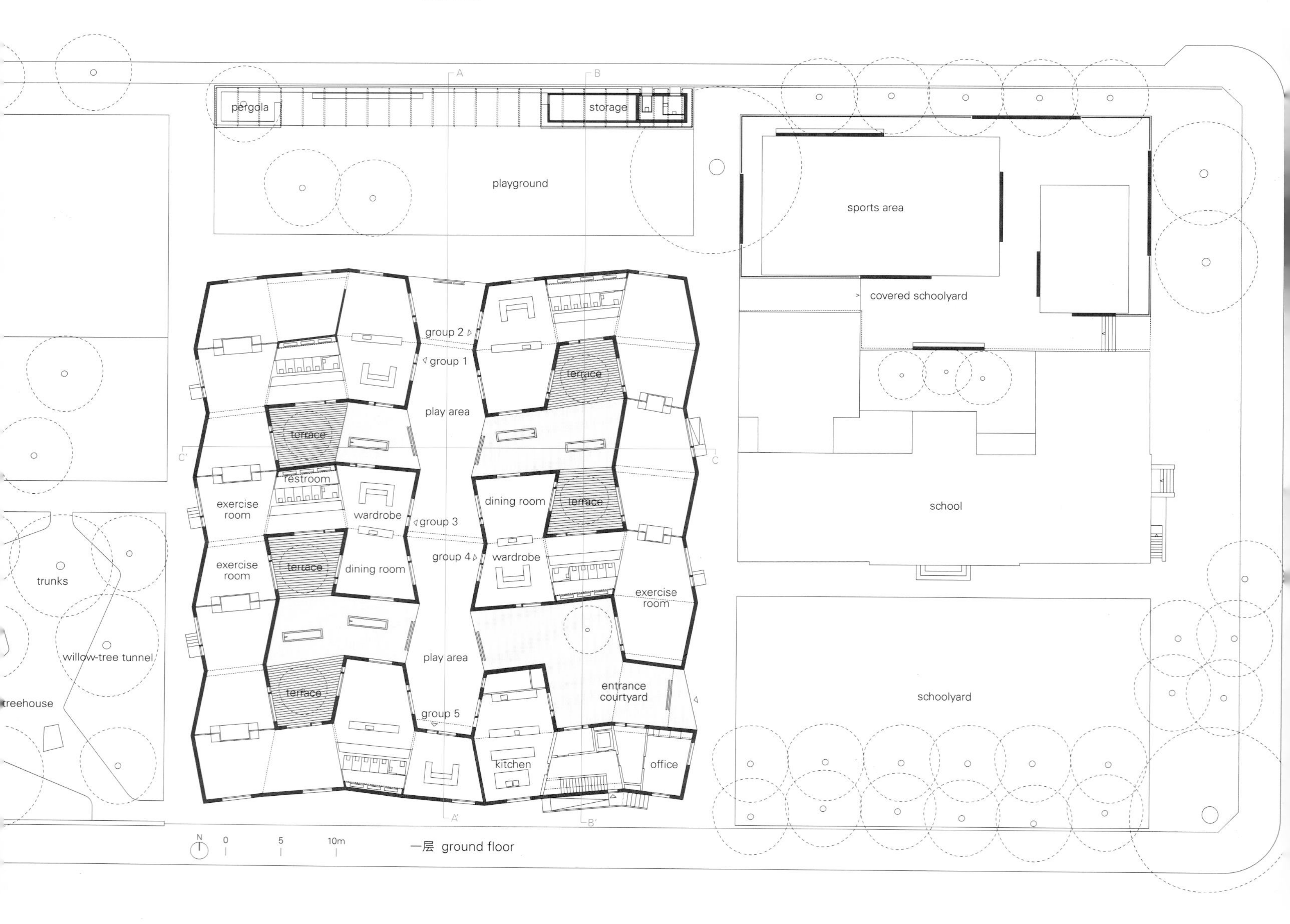

一层 ground floor

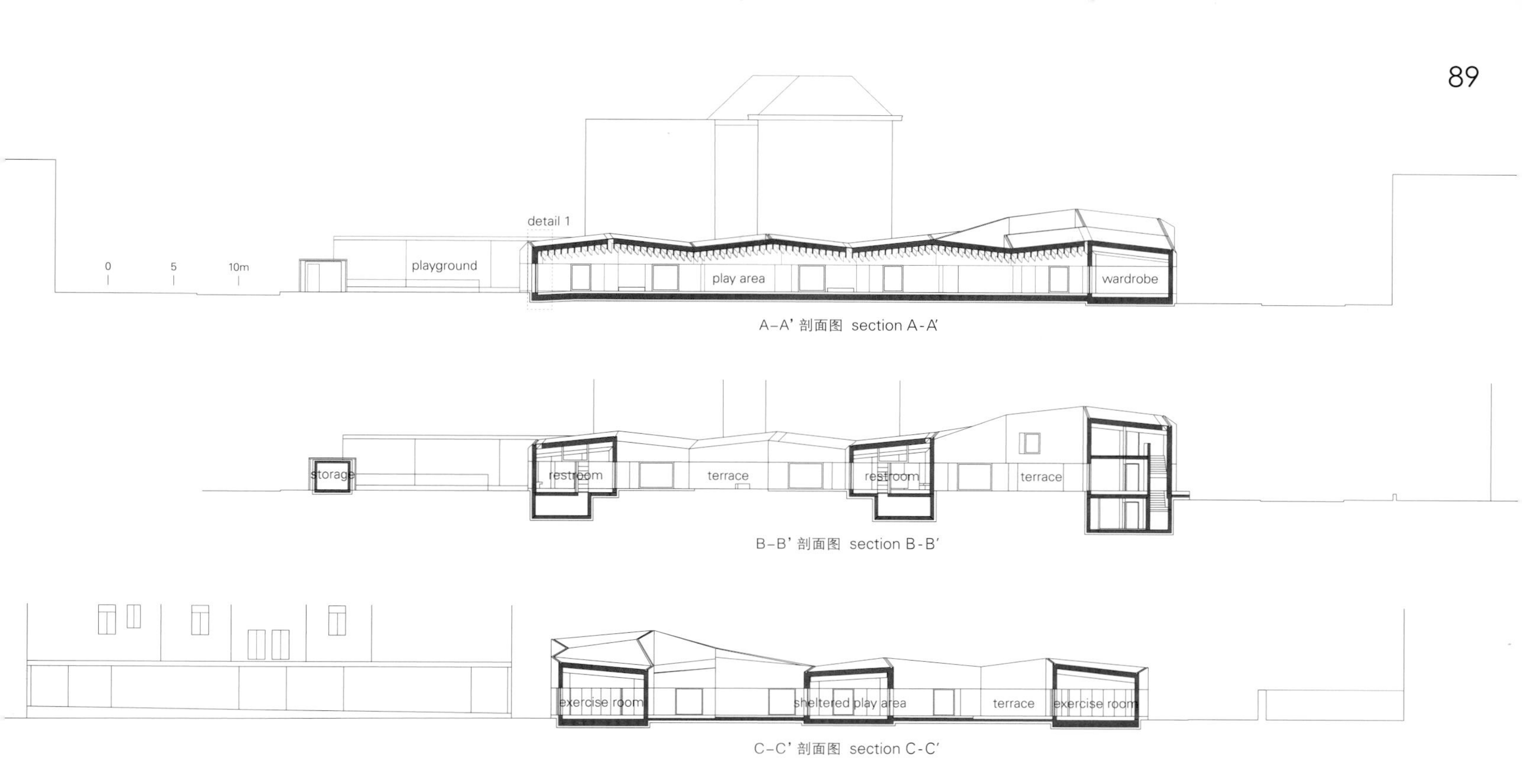

A–A' 剖面图 section A-A'

B–B' 剖面图 section B-B'

C–C' 剖面图 section C-C'

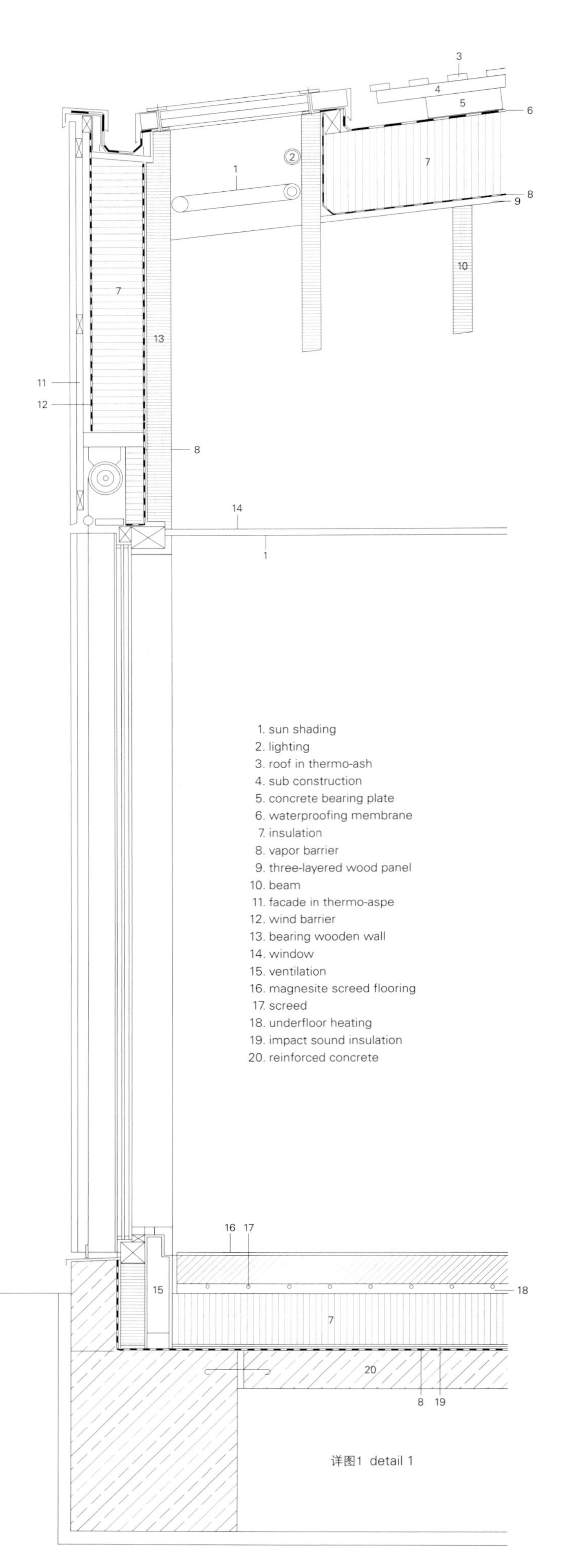
1. sun shading
2. lighting
3. roof in thermo-ash
4. sub construction
5. concrete bearing plate
6. waterproofing membrane
7. insulation
8. vapor barrier
9. three-layered wood panel
10. beam
11. facade in thermo-aspe
12. wind barrier
13. bearing wooden wall
14. window
15. ventilation
16. magnesite screed flooring
17. screed
18. underfloor heating
19. impact sound insulation
20. reinforced concrete

详图1 detail 1

布尔镇幼儿园

Dominique Coulon & Associes

该建筑成为阿尔萨斯山谷里的一座小村庄的入口标志。一座 14 世纪的城堡位于附近的山坡上。幼儿园建筑与这座防御城堡的正交几何外形遥相呼应。幼儿园周围建有带洞口的城堡式围墙，来保护在操场玩耍的孩子。这种空间布局为望向孚日山的人们提供了圆形的视野范围。

该建筑被设计为规整的矩形平面，布局为一些连续的皇冠形体量，这些皇冠形体量内容纳了项目的不同功能。体量的层次使整体建筑具有一定的深度。建筑中心位置的高度增加了一倍，在自然光线的照射下

好似一个万花筒，色彩丰富。这些立方体的表面涂有不同的色彩，从粉红色到红色。亚光面和光面的表面处理效果产生一定的共鸣，使空间更加丰富，也更加精致。

不同的体量层次之间具有不同的透明度，使建筑的深度显而易见。通过天窗照射进来的充足光线能够穿过整个内部空间。该建筑就好像是这片景观内零散碎片堆砌的一个大型建筑物，实心与空心的表现形式使人们想起了乐高积木搭建起来的城堡，建筑周围还有68棵苹果树，使人们想起了当地的农业景观。

Nursery in Buhl

The building marks the entrance to a small village nestling in a valley in Alsace. A 14th-century castle dominates the site from the nearby hillside. The day nursery echoes the orthonormal geometry of the fortified castle. A perimeter wall with openings like on a castle wall protects the children's play-

grounds. This spatial arrangement offers views of the rounded outlines of the Vosges mountains.

The principle of the strictly rectangular plan is an arrangement of successive crowns containing various elements of the project. These layers give depth to the project overall. The heart of the building is formed by a central space which emerges at double heights and plays with natural light like a kaleidoscope. These cubic volumes condense a host of faces ranging in color from pink to red. The matte and shiny colors resonate, shaping the space to make it richer and more subtle. The multiple transparencies installed between different layers give an indication of the depth of the building. There is abundant natural light throughout, captured by skylights emerging from the overall volumes. The building appears in the landscape like a fragmented monolith where the play of solids and hollows is reminiscent of something like a Lego model of a castle. The building is surrounded by sixty-eight apple trees which hark back to the local agricultural landscape.

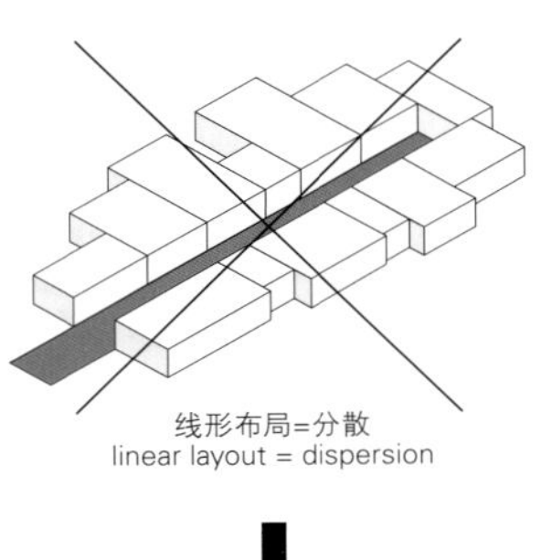

线形布局=分散
linear layout = dispersion

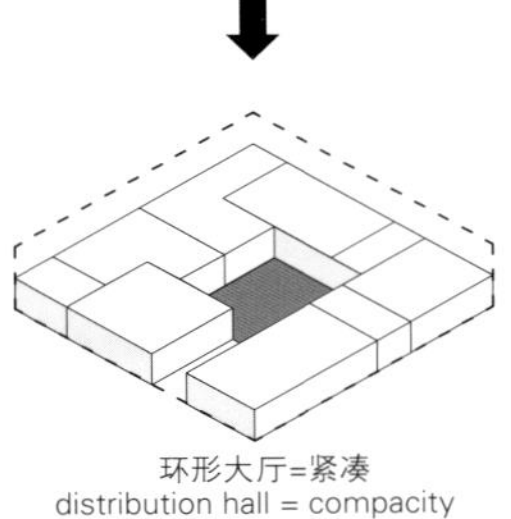

环形大厅=紧凑
distribution hall = compacity

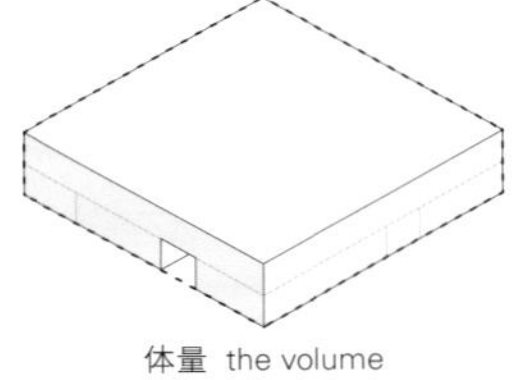

体量 the volume

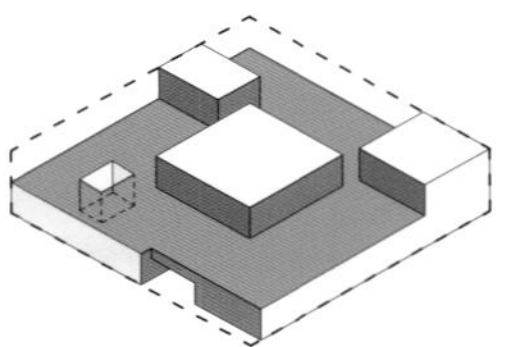

切割的体量 the carved volume

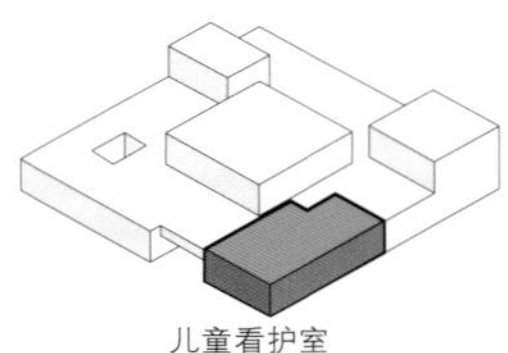

儿童看护室
nursery assistant room

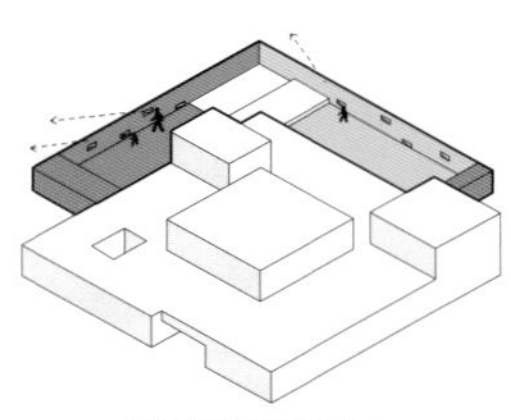

受保护的室外空间
protected outdoor spaces

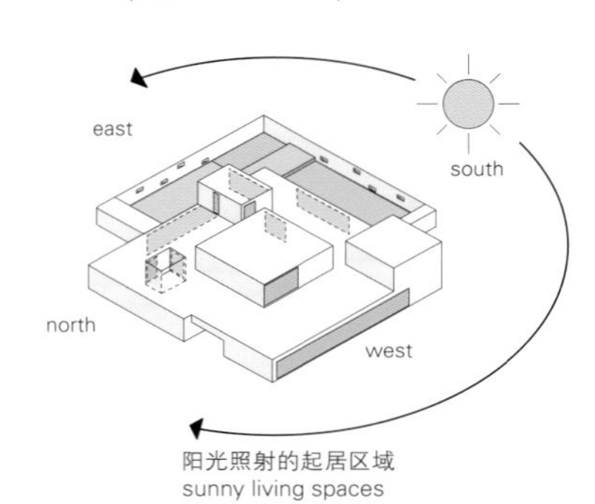

阳光照射的起居区域
sunny living spaces

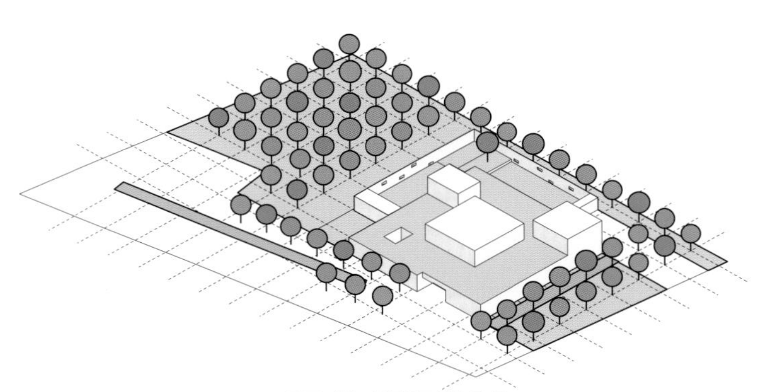

果园、灌溉沟渠以及绿化屋顶
orchard, drainage ditches and green roofs

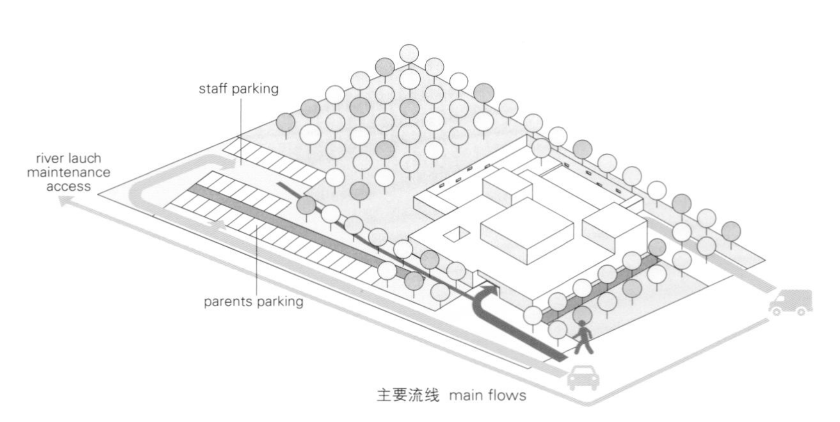

主要流线 main flows

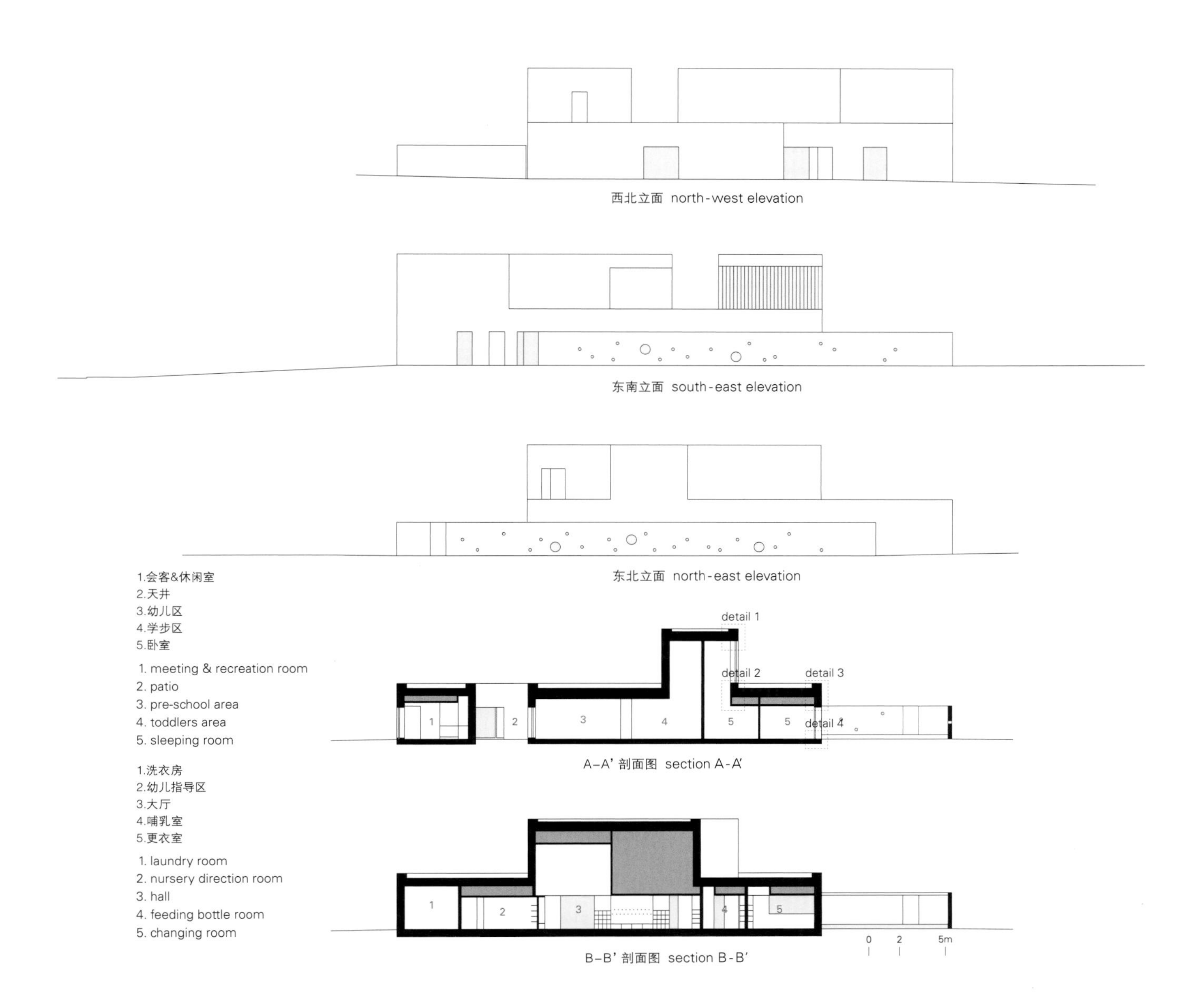

西北立面 north-west elevation

东南立面 south-east elevation

东北立面 north-east elevation

A–A' 剖面图 section A-A'

B–B' 剖面图 section B-B'

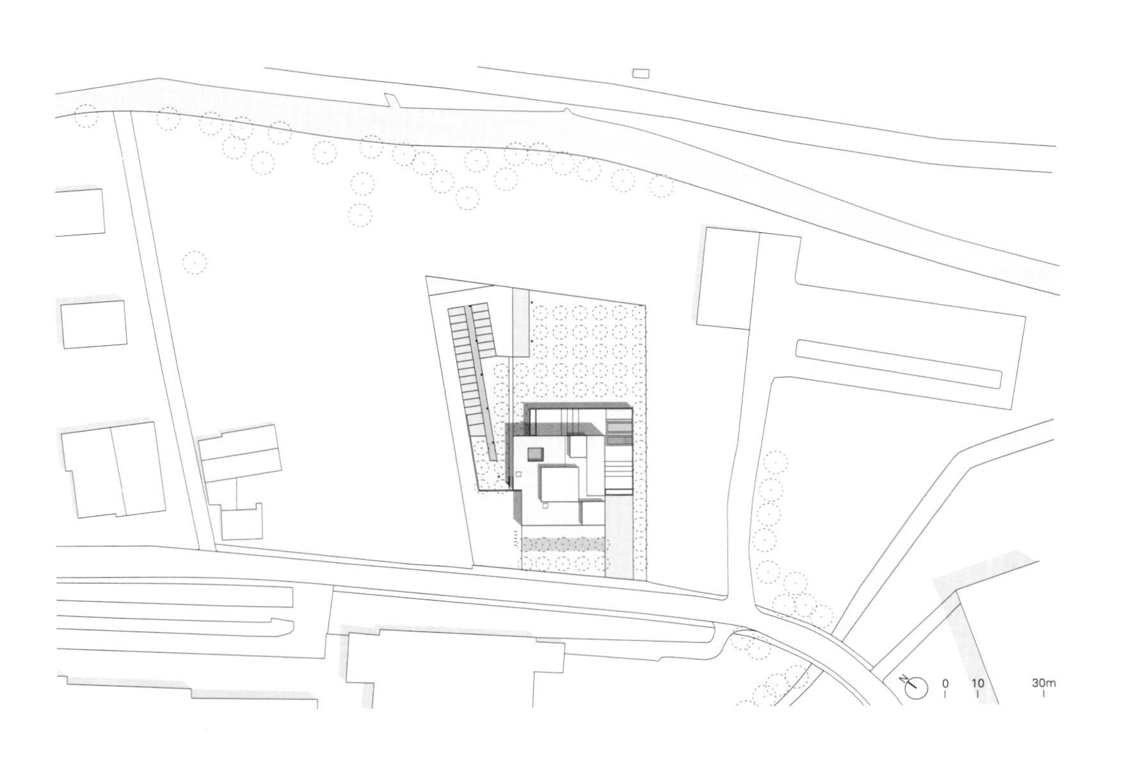

项目名称：Nursery in Buhl / 地点：14 rue de la Fabrique, 68530, Buhl, France
建筑师：Dominique Coulon, David Romero-Uzeda, Olivier Nicollas_Dominique Coulon & associés
助理：Javier Gigosos Ruipérez, Diego Bastos-Romero, Gautier Duthoit
场地施工监督：David Romero-Uzeda / 结构工程师：Batiserf
电气工程师：BET G.Jost / 机械&水暖工程师：Solares Bauen
估价：E3 économie /声效：Euro sound project /厨房设计：Ecotral / 景观设计：Philippe Obliger
功能：Nursery, Day nursery and Nursery assistants relay for 40 children
用地面积：5,171m² / 总建筑面积：706m² / 有效楼层面积：763m²
客户：CC Région de Guebwiller / 造价：EUR 1,700,000
竞标时间：2013.3 / 施工时间：2014.2—2015.6
摄影师：©Eugeni Pons (courtesy of the architect) (except as noted)

SALLE
D'ACTIVITES
PSYCHOMOTRICITE

1.幼儿指导区
2.儿童看护员指导区
3.卧室
4.婴儿室
5.学步区
6.幼儿区
7.大厅
8.天井
9.会客室
10.心理咨询室
11.餐厅&活动室
12.厨房
13.洗衣房
14.会客室&休息室
15.婴儿游乐区
16.学步儿童/学前儿童游乐区
17.蔬菜园

1. nursery direction room
2. nursery assistants direction room
3. sleeping room
4. babies area
5. toddlers area
6. pre-school area
7. hall
8. patio
9. meeting room
10. psychomotricity room
11. dinning and activities room
12. kitchen
13. laundry room
14. meeting & recreation room
15. babies playground
16. toddlers / pre-school playground
17. vegetable garden

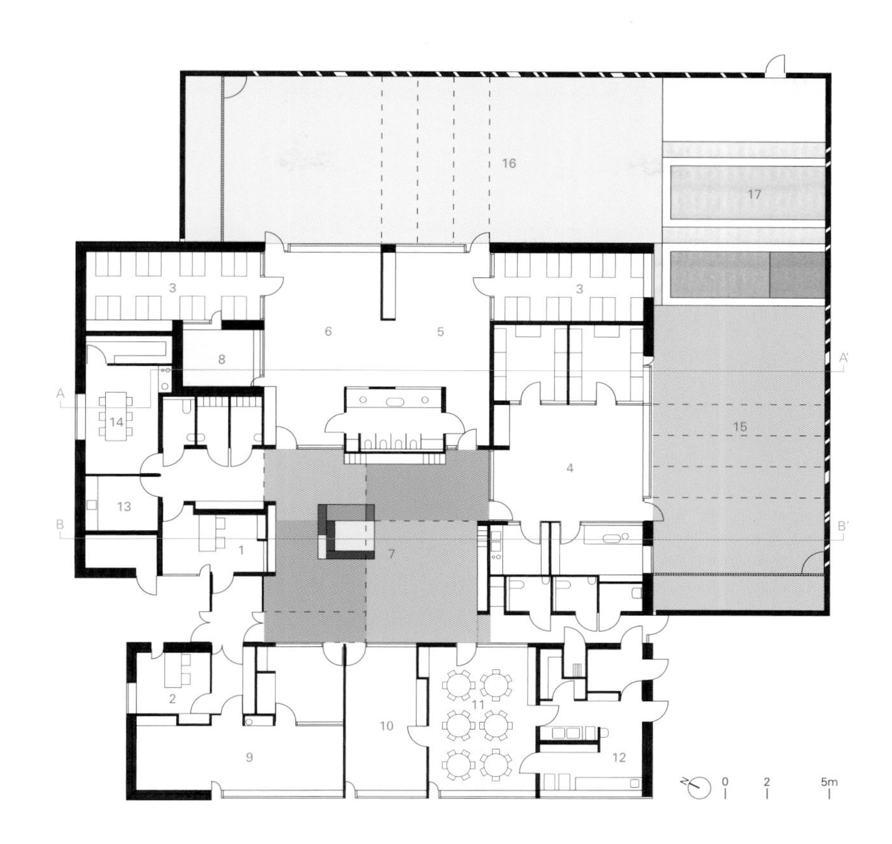

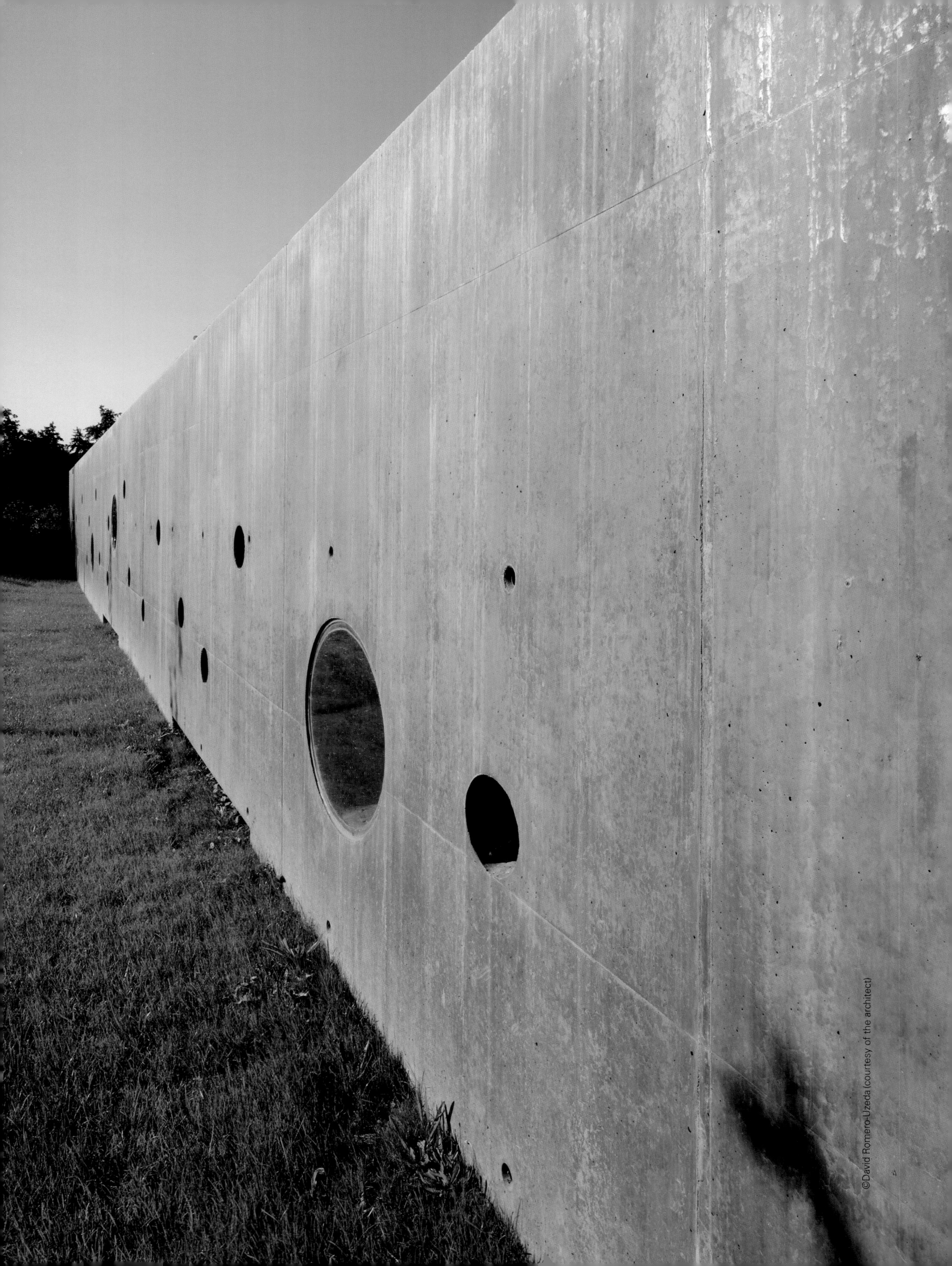

©David Romero-Uzeda (courtesy of the architect)

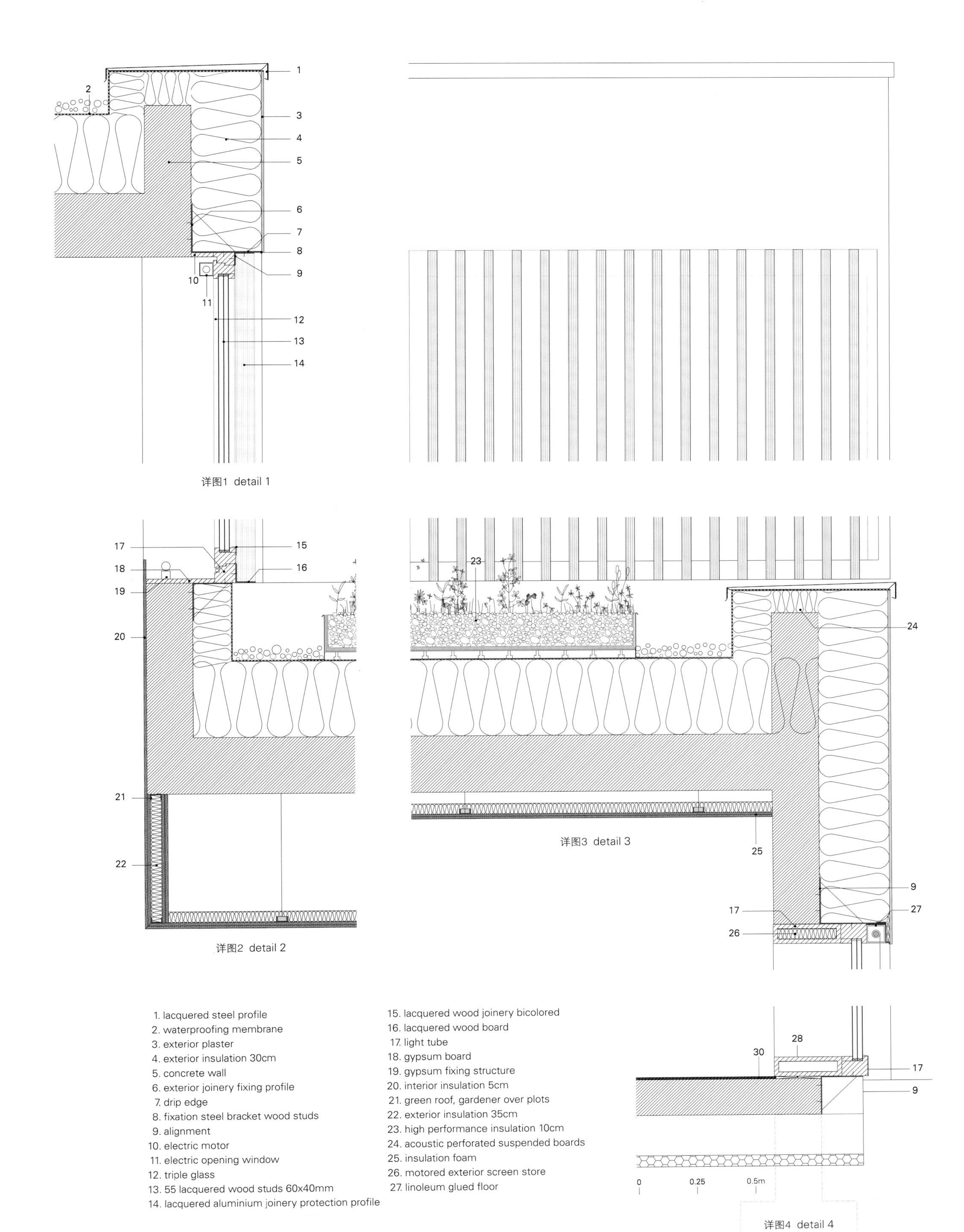

详图1 detail 1

详图2 detail 2

详图3 detail 3

详图4 detail 4

斯塔比奥幼儿园

studio we architetti

这座幼儿园独立伫立在一片绿地上。新建筑仅通过道路和墙体来和景观连接，为场地内的参考点营造了微妙但浓重的质感。

建筑紧凑的结构可以通过几个基本元素来表达。入口上方的波浪形屋顶和天窗定义了建筑正前面的立面轮廓。该幼儿园由四间教室组成，室外空间和室内房间之间还建立了对话。每间教室都具有这种空间关系。

这种设计手法可以使孩子们在这里以一个自然且有趣的方式来体验空间，并且帮助他们成长。不同建筑背景的主题都有所限定，来引导孩子们在自由与隐私之间探索。空间举办的不同活动可交替体现孩子的外向及内向表现瞬间。水平和垂直的连接也使孩子们能从每个单元来感知整个设施。

身处这座幼儿园如同位于一座小镇之中，不同的大小和形状的空间、光、材料和颜色决定了空间的丰富性和差异性。

Kindergarten in Stabio

The kindergarten is a solitary building on a green field. Only paths and walls connect the new building to landscape in order to create a subtle and strong texture of reference points within the site.

The architectural expression of the compact structure can be detected through few basic elements. The shape of the roof with its geometric waves over the entrance and its skylights

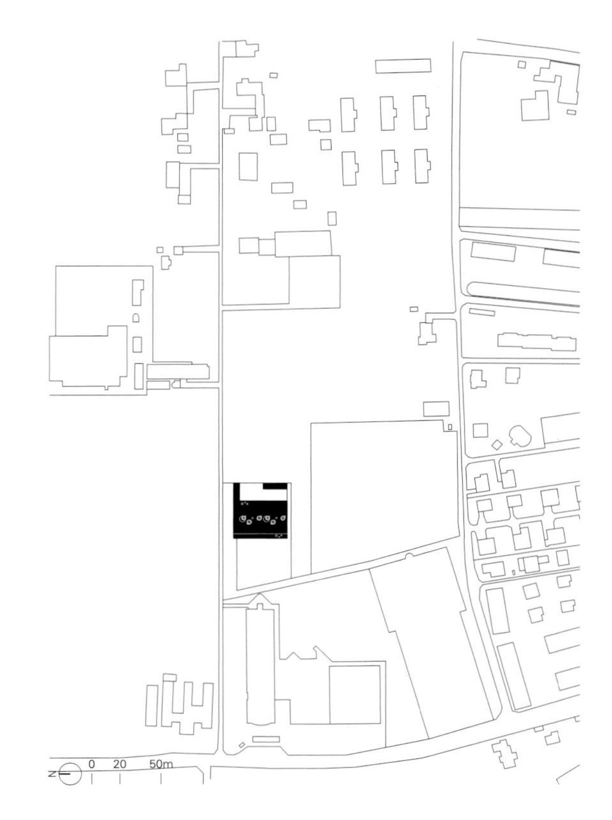

项目名称：Kindergarten in Stabio
地点：Via Luvee 15, 6855 Stabio, Switzerland
建筑师：Felix Wettstein _ studio we architetti
结构工程师：Edy Toscano Engineering & Consulting, Rivera
电气工程师：Scherler SA, Lugano Switzerland
用地面积：3,353m² / 总建筑面积：1,028m² / 有效楼层面积：1,848m²
设计时间：2008 / 竣工时间：2013
摄影师：©Alexandre Zveiger (courtesy of the architect)

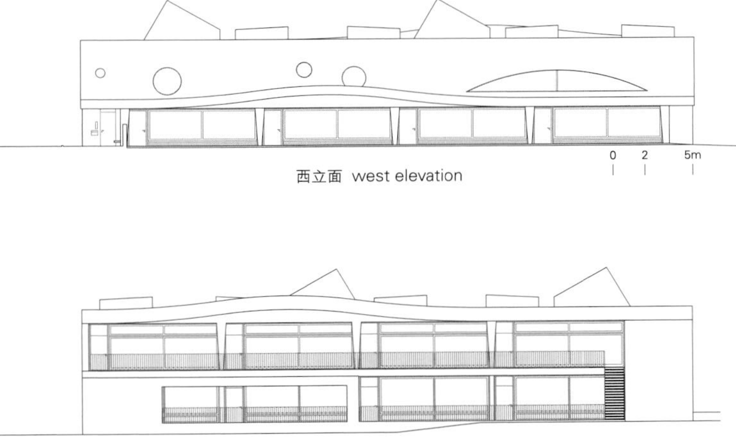

西立面 west elevation

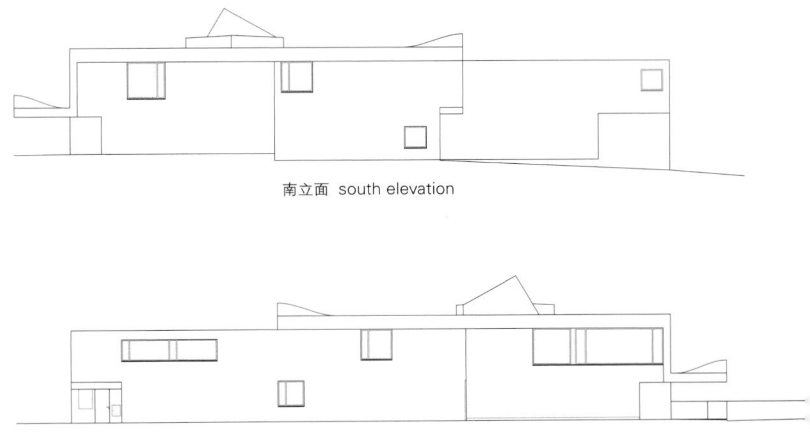

南立面 south elevation

东立面 east elevation

北立面 north elevation

define the elevation of the front facade of the building. The kindergarten is composed by four classes and it creates the dialogue between interior rooms and exterior spaces. The typology of each class indicates this relationship.

Children could experiment spaces in a natural and funny way that contributes to their growing. The theme of setting limits in its different varieties, guiding children in their search for freedom as well as privacy. Extroverted and introverted moments alternate in a series defined by the different activities. Horizontal and vertical connections allow the perception of the overall facility from each unit.

As in a small town, spaces of different size and shape, light, materials and colors determine the richness and the different identities of the space. studio we architetti

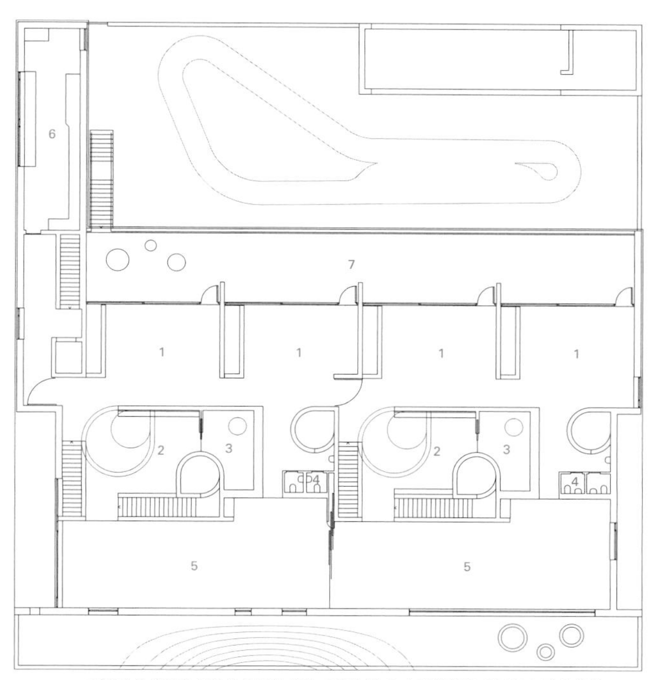

1.餐厅 2.校长办公室 3.员工办公室 4.卫生间 5.心理咨询室 6.厨房 7.室内操场
1. dining room 2. head teacher office 3. staff room 4. toilet
5. psychomotor classroom 6. kitchen 7. sheltered playground
二层 first floor

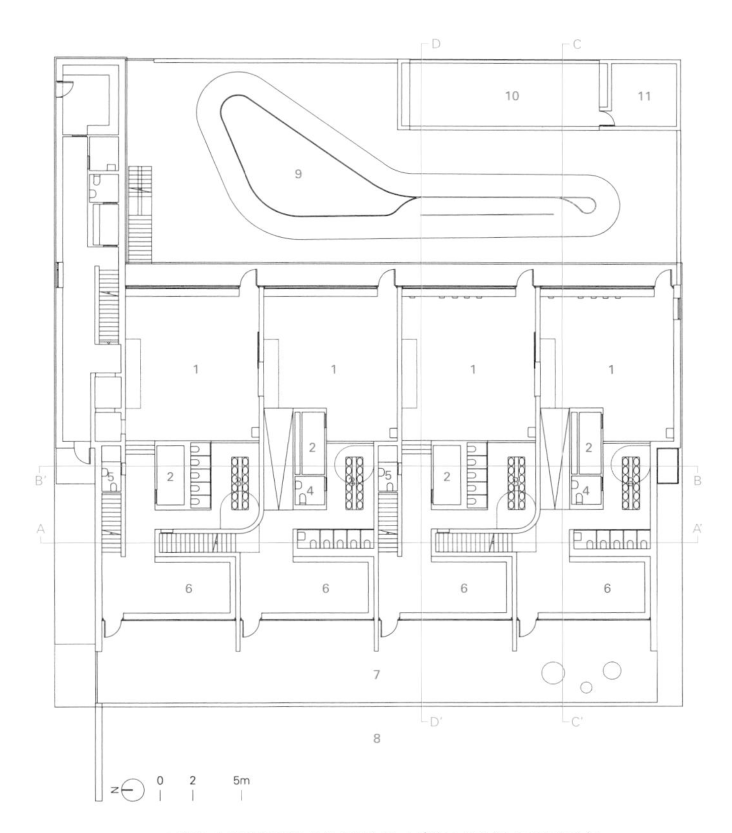

1.教室 2.教室储藏室 3.学生卫生间 4.残疾人卫生间 5.员工卫生间
6.衣橱 7.入口门廊 8.入口广场 9.学校操场 10.室内操场 11.花园储藏室
1. classroom 2. classroom storage 3. toilet pupils 4. toilet for the disabled
5. staff toilet 6. wardrobe 7. access porch 8. access plaza 9. school playgournd
10. sheltered playground 11. garden storage
一层 ground floor

A–A' 剖面图 section A-A'

B–B' 剖面图 section B-B'

C–C' 剖面图 section C-C'

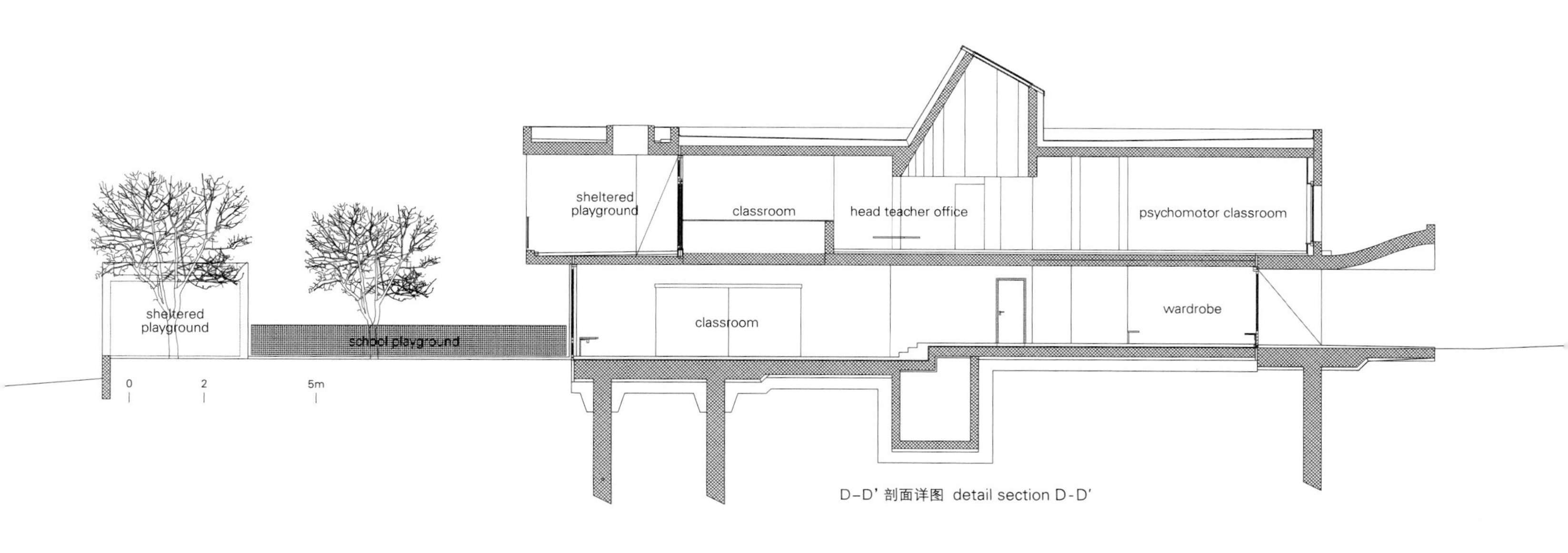

D-D' 剖面详图 detail section D-D'

该学校由八间幼儿园教室和十间小学教室组成，此外还设有一间学校自助餐厅和一个休闲中心。该建筑有三个楼层和一个地下层，地下层内还设有服务空间。秉承着场地内的各个单元都设有相同数量的通道和其他连接的原则，施工场地被分为六个部分：三处户外区（入口、幼儿园操场和小学操场）和三处室内区（幼儿园、小学与带休闲中心的自助餐厅）。这种分割形成了一个三叶草形的布局，这种布局除了设有人们聚会的走廊外，还在内部和外部之间形成一个较好的连接。

在建筑内部，考虑到幼儿园和小学的教学差异，我们开发了两处不同的空间，就像人们的左脑和右脑一样。我们的大脑半球是对称的，但并不完全相同，因为他们具有各自独特的重要功能。所以这座建筑只是在外观上是对称的。

幼儿园教室由三个不同直径和高度的圆形空间组成。小学教室则是方形的，一侧完全是玻璃墙面。圆形空间之间的孔隙设为存储区，来留出更多的空间设置教室。

多功能的幼儿园空间是圆形的，天花板则是半球形，而小学教室空间是方形的，天花板是锥形的。这两处大型空间都面向操场开放，且顶部设有照明设施。同时这两处空间面向走廊开放，有助于室内自然采光，且利用室内外之间的透明度，易于引导人们规划流线。

我们的一个设计重点是使空间通畅，易于引导。从入口处开始，大厅的三叶草形结构使访客能够直接看到两个操场。在大厅里，三条通道在此汇聚，幼儿园在右侧，小学在左边，休闲中心就在围绕电梯设置的大楼梯的前面。

大厅纵跨建筑的整个高度，被顶部的照明设备照亮。悬挂的两个半球结构使这处大型上空空间富有生气：每个半球结构都是属于幼儿园和小学图书馆的阅读空间。

考虑到建筑的三叶草形结构，立面的施工过程并非容易。为了避免形成单调而固定的立面，我们设计了一个木质覆层系统，当游客围绕着建筑移动时，板条发生了变化。立面的下半部为橙色，而上半部涂成

圣丹尼斯幼儿园和小学

Paul Le Quernec

了苹果绿色，且前侧板条的表面纹理粗糙，未经加工。这种设计手法使建筑的正面给人一种非彩色的视觉效果，但是当我们从倾斜的角度来观看时，这种视觉效果便会减弱，从而呈现为彩色。当我们走进校园时，主立面为绿色，而走出校园时，则为橙色。这种效果是一种开始便形成的视觉错觉，对于训练人的思维起到了很大的作用。这个项目设计的每一个选择都考虑到了对孩子们心理－生理发展的影响。

Nursery and Primary School in Saint-Denis

The building is composed of a nursery with eight classrooms, a primary school with ten classrooms, a school cafeteria and a recreation center. The building has three levels and a basement which contains a service room. For an equal allocation of the access and connections to the different units, the construction site has been divided into six parts: three outdoor (entrance, nursery playground and primary playground) and three indoor (nursery school, primary school and school cafeteria with recreation center). From this division results a clover-shaped building that offers, in addition to convergent and economic corridors, a very good connection between the inside and the outside.

Inside the building, considering the teaching difference between the nursery school and the primary school, we designed two spaces as different as the right brain can be from the left. Our brain hemispheres are symmetrical, but not

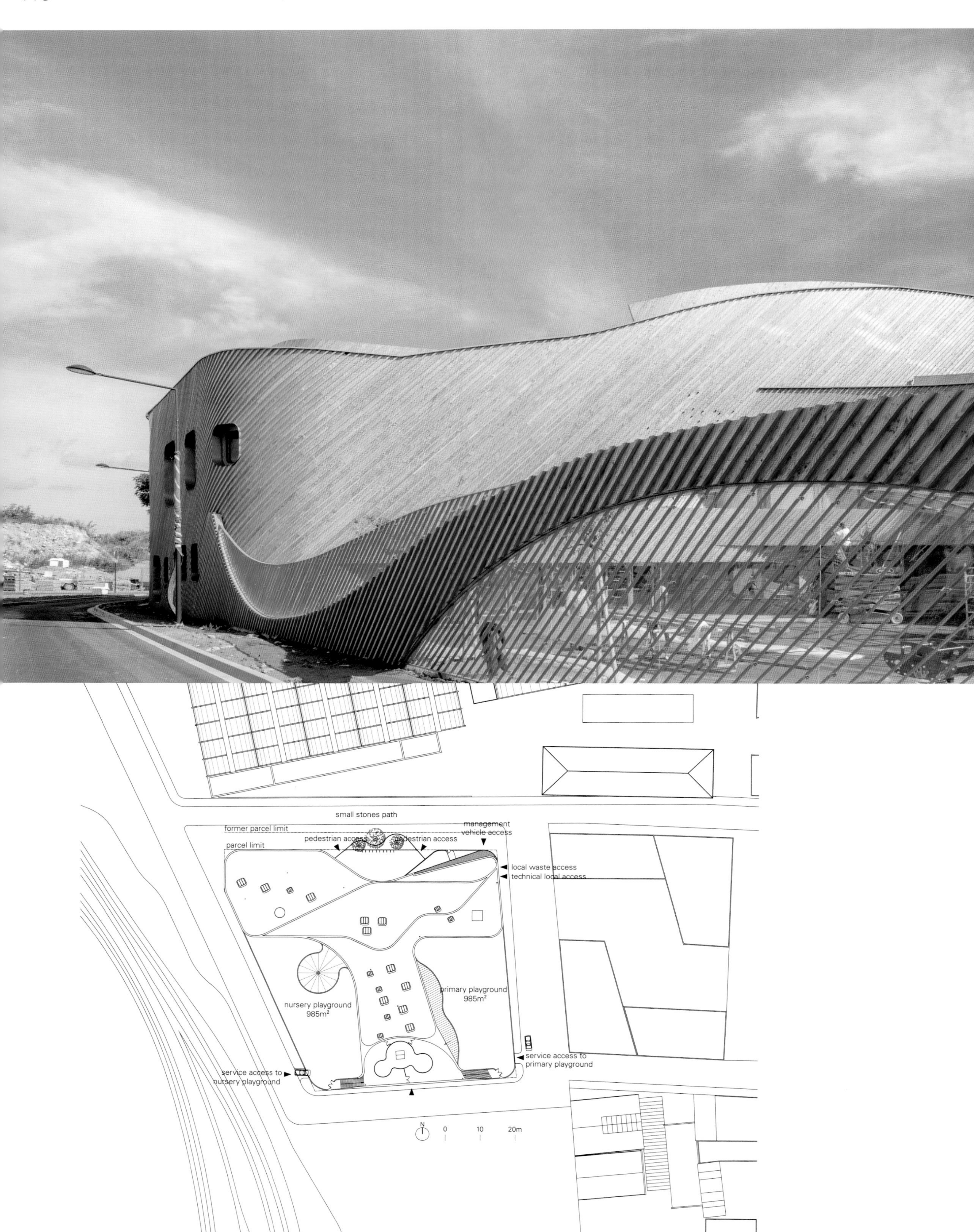
small stones path
former parcel limit
pedestrian access
pedestrian access
management vehicle access
parcel limit
local waste access
technical local access
nursery playground 985m²
primary playground 985m²
service access to primary playground
service access to nursery playground
N
0
10
20m

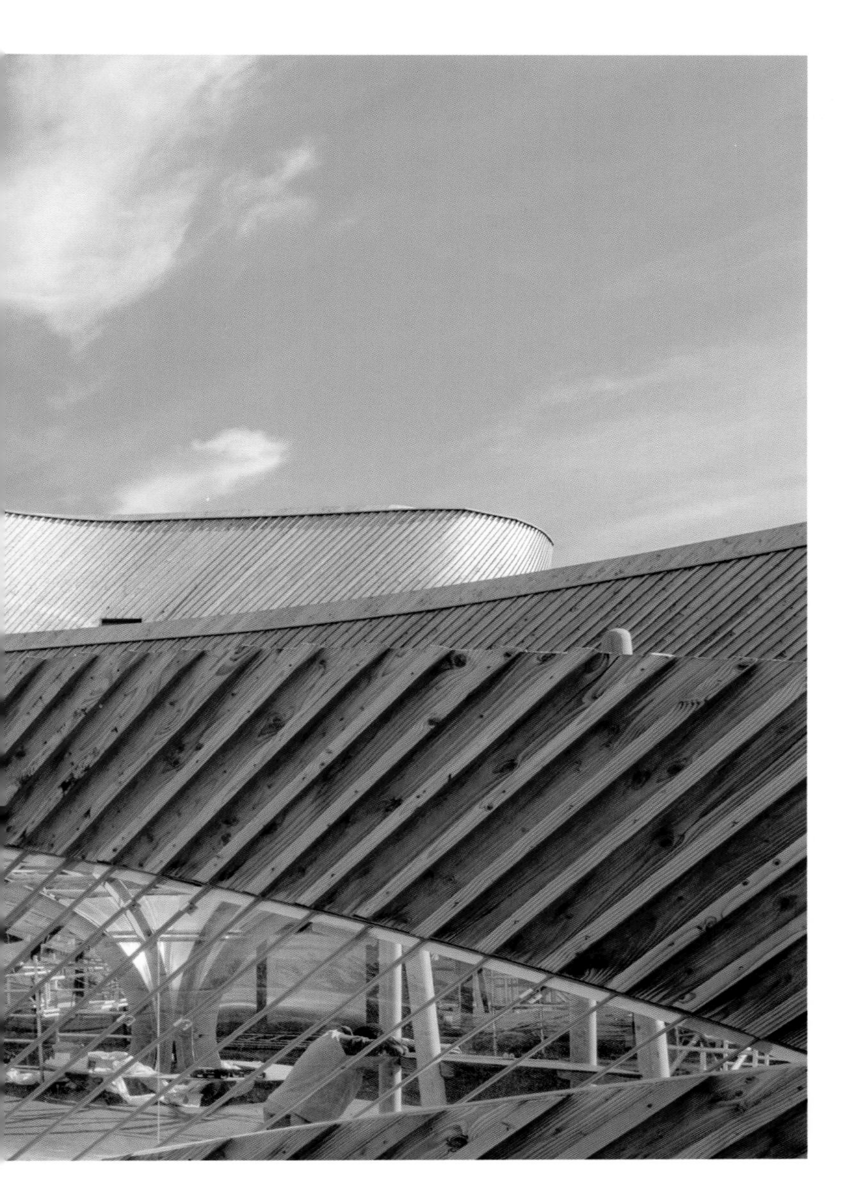

项目名称：Nursery and Primary School in Saint-Denis (93200)
地点：1 Bis chemin des petits cailloux, 93200 Saint-Denis, France
建筑师：Paul Le Quernec
业主：City of Saint Denis, France
功能：nursery school, primary school, school cafeteria, recreation centre
用地面积：4,842m² / 总建筑面积：4,800m² / 有效楼层面积：4,600m²
造价：EUR 12,000,000
设计时间：2012 / 施工时间：2015 / 竣工时间：2015.8
摄影师：©11h45 (courtesy of the architect)

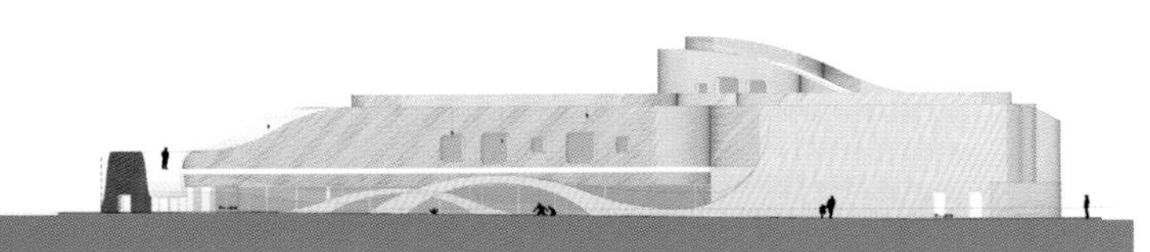

东立面 east elevation

北立面 north elevation

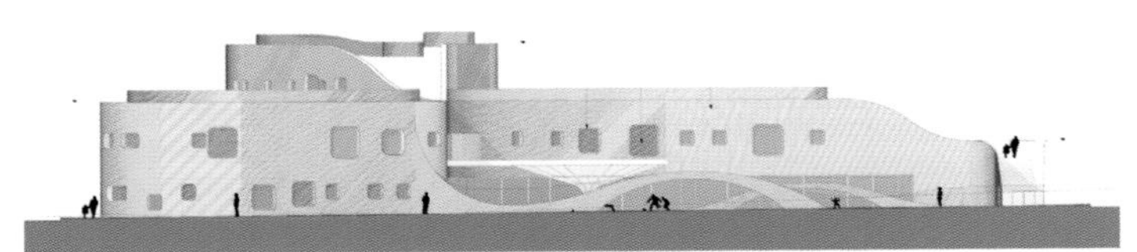

西立面 west elevation

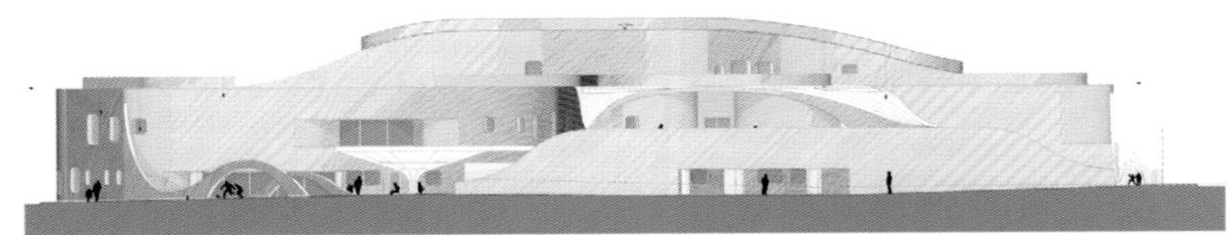

南立面 south elevation

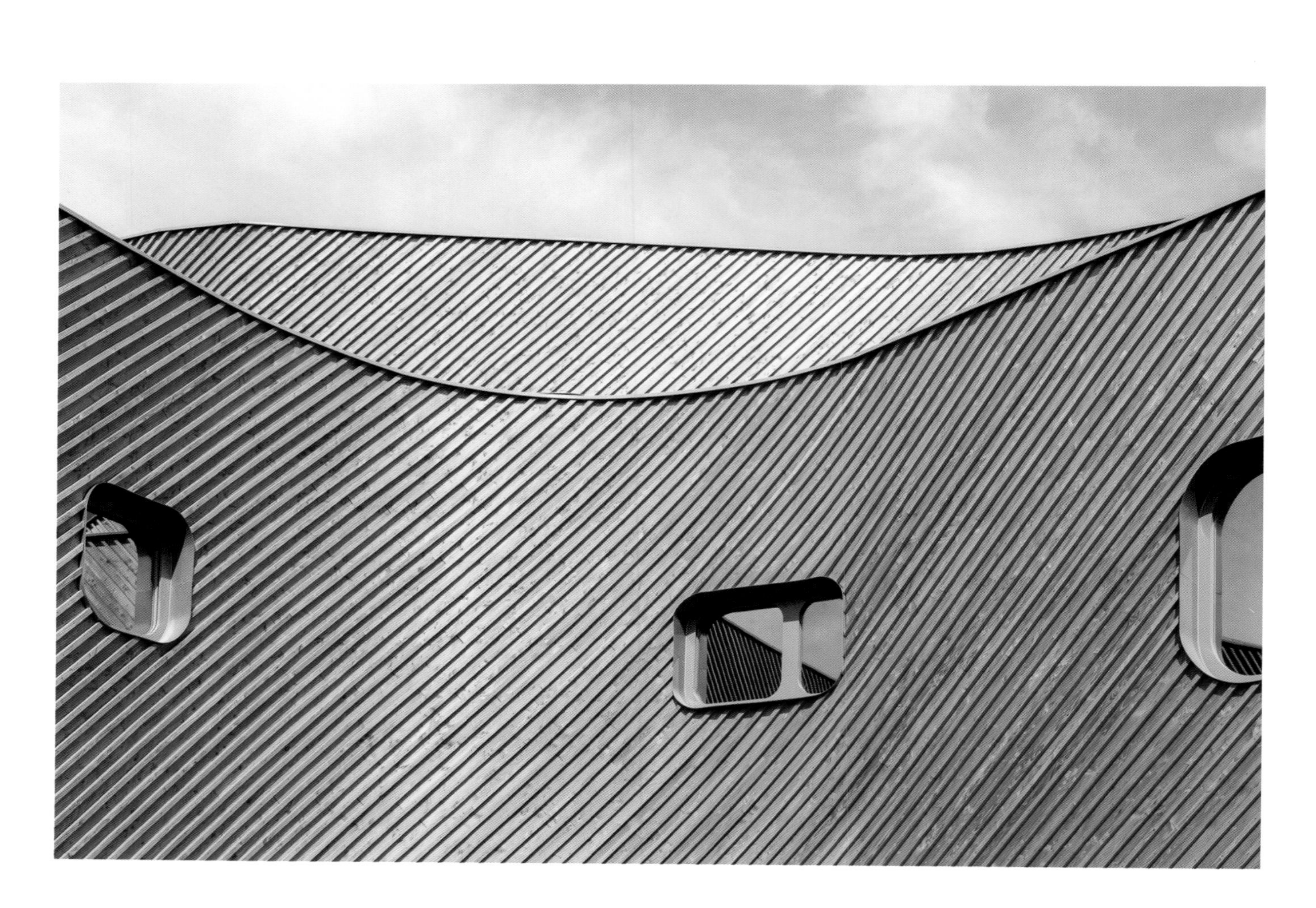

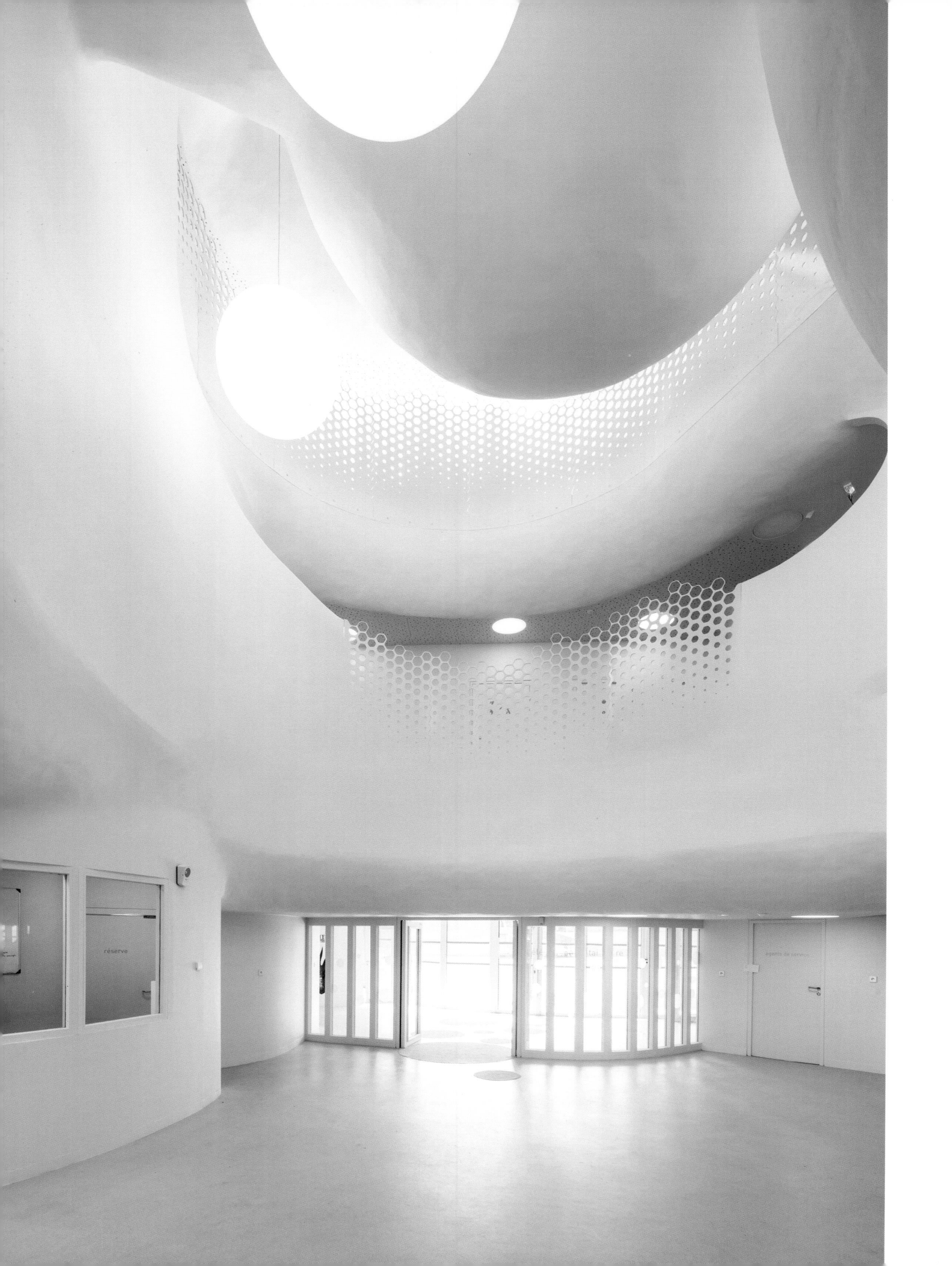
réserve

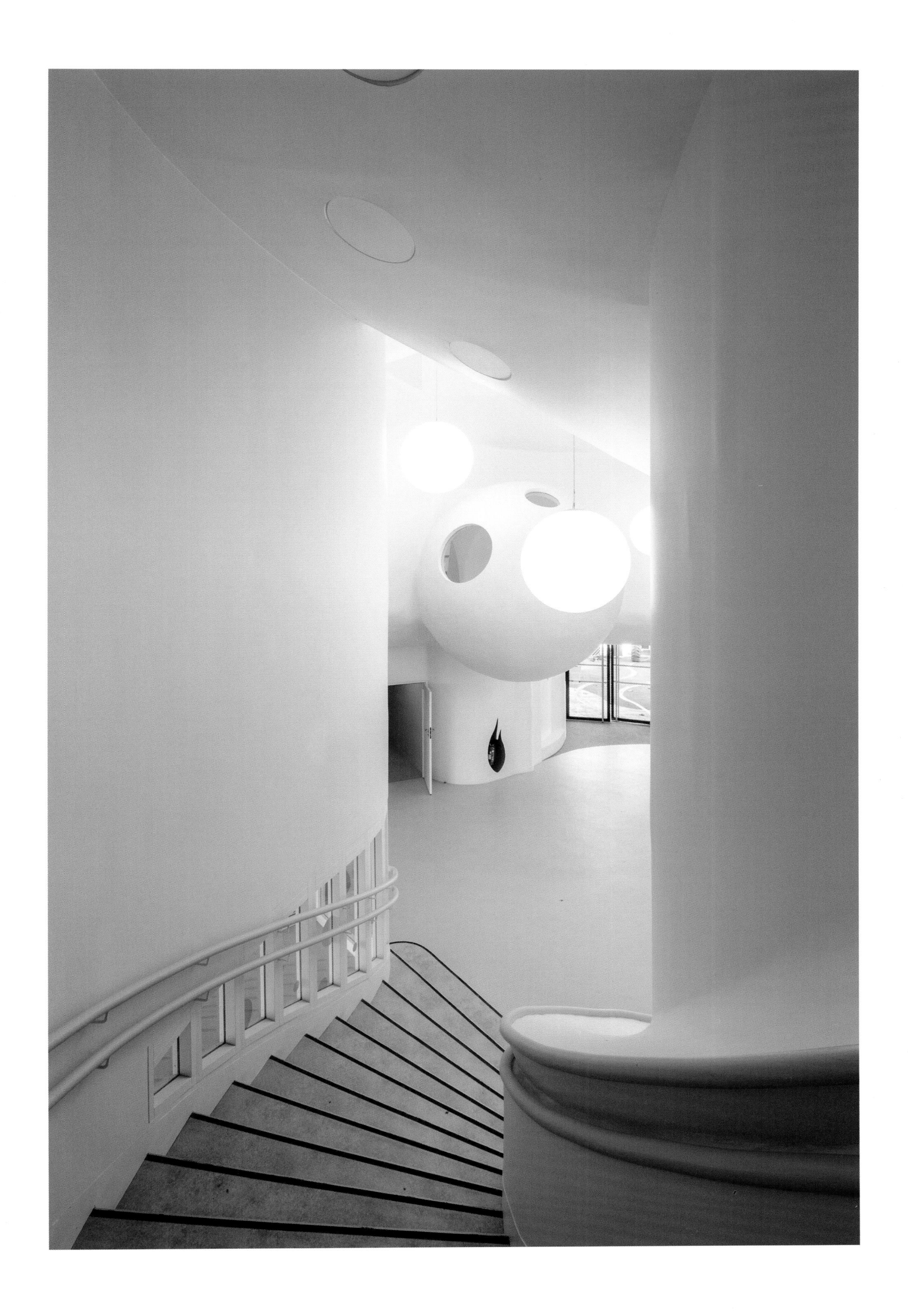

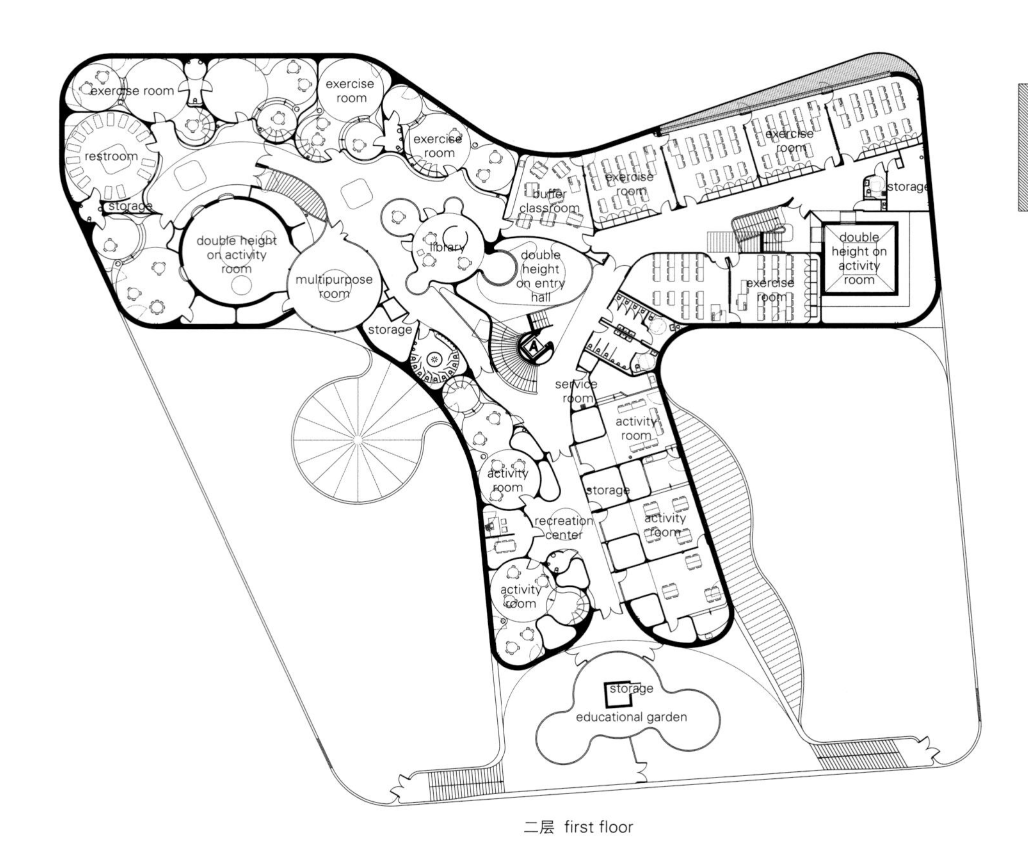

二层 first floor

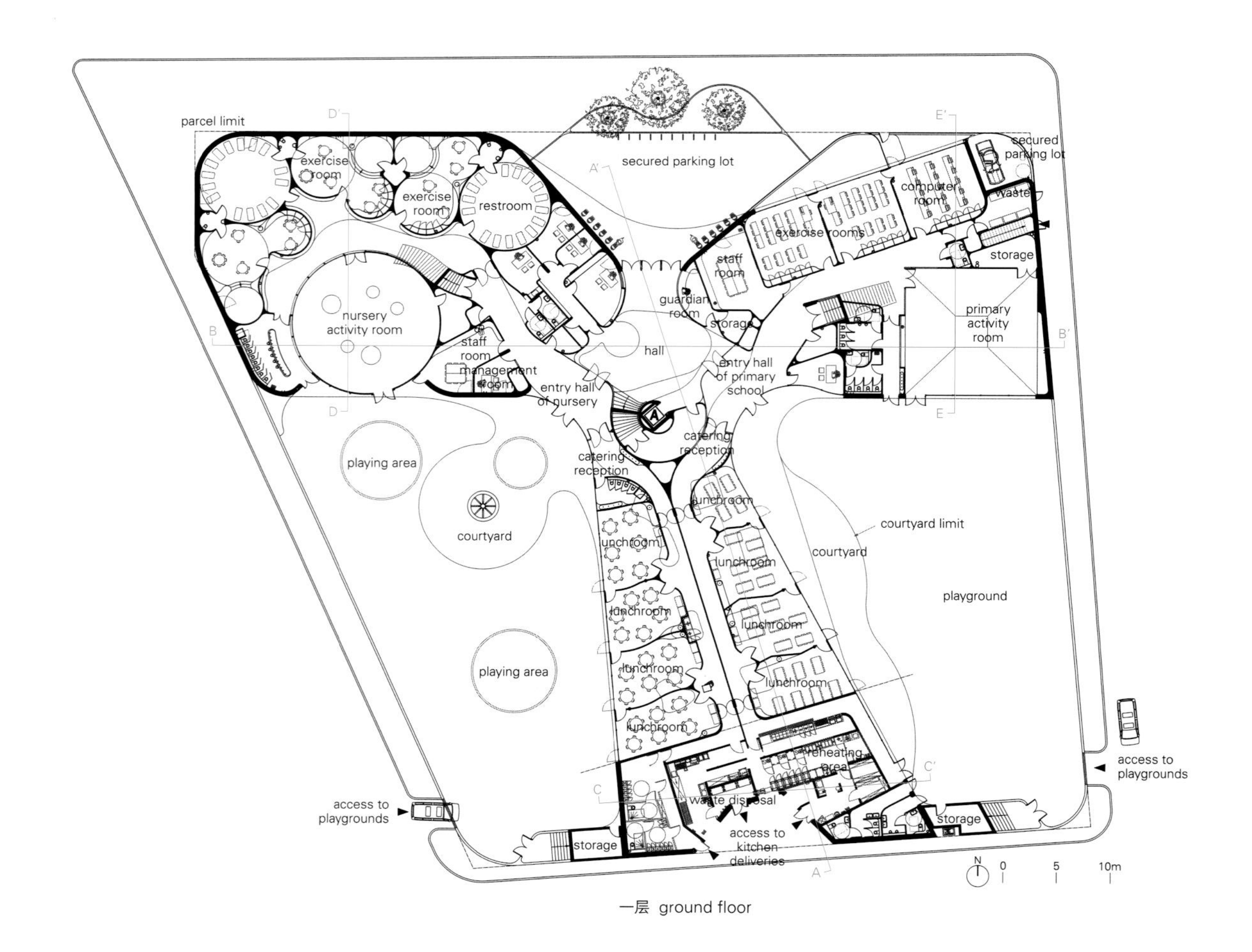

一层 ground floor

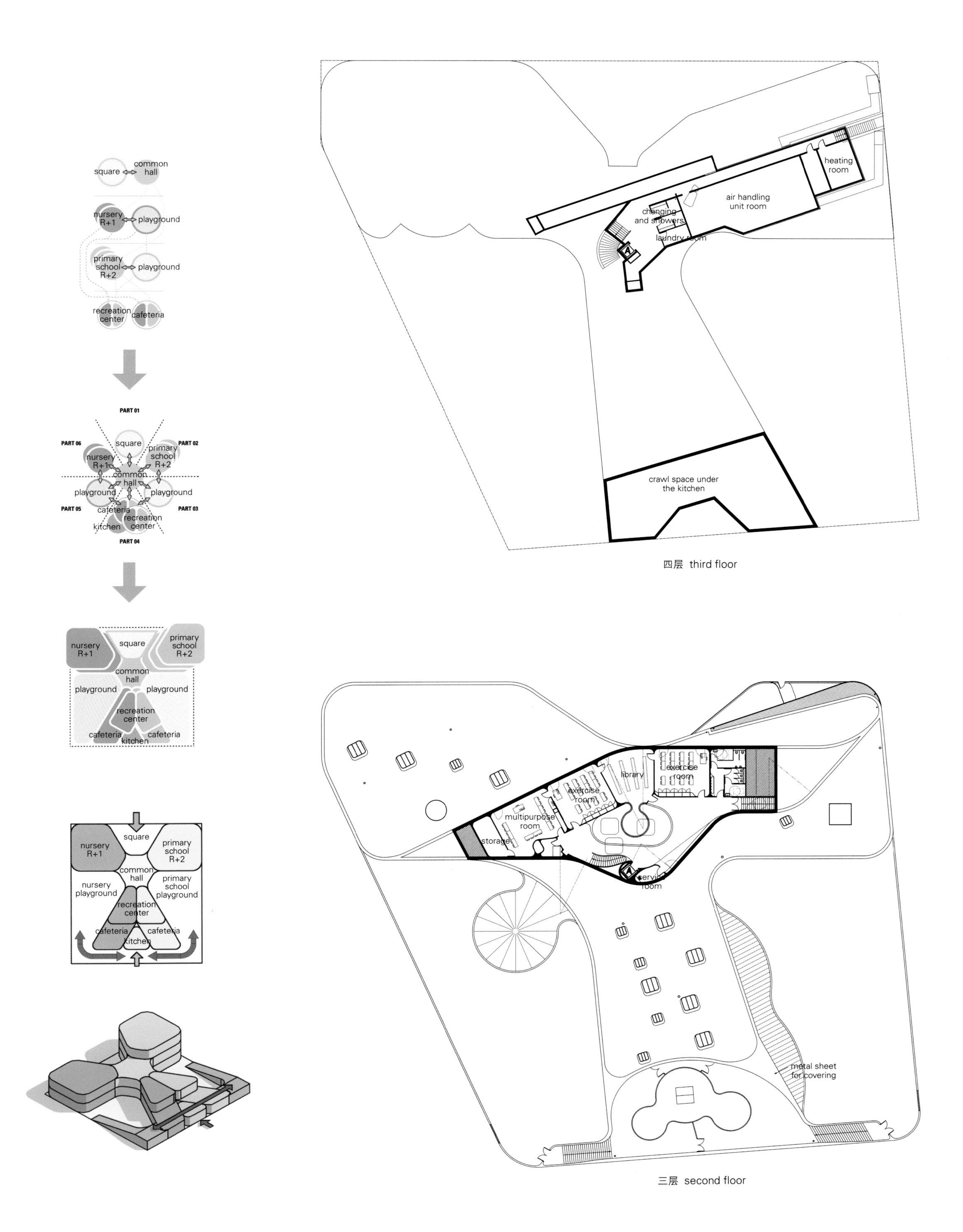

四层 third floor

三层 second floor

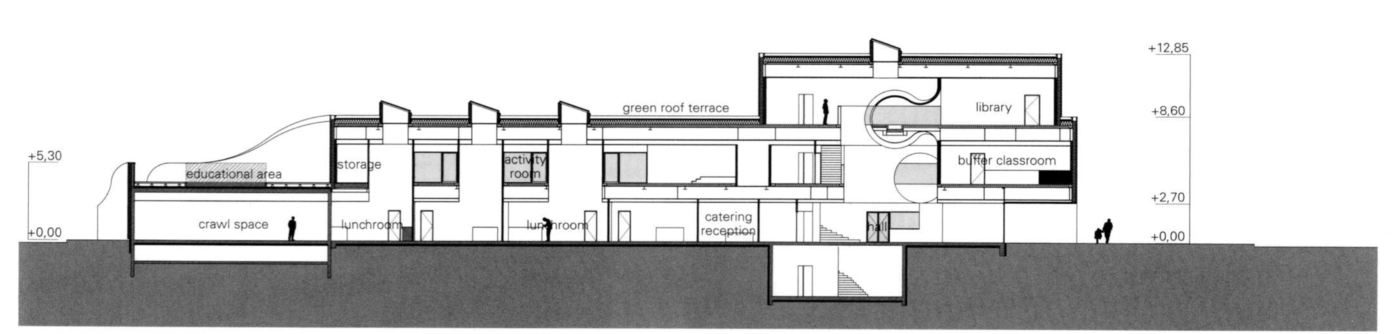

A–A' 剖面图 section A-A'

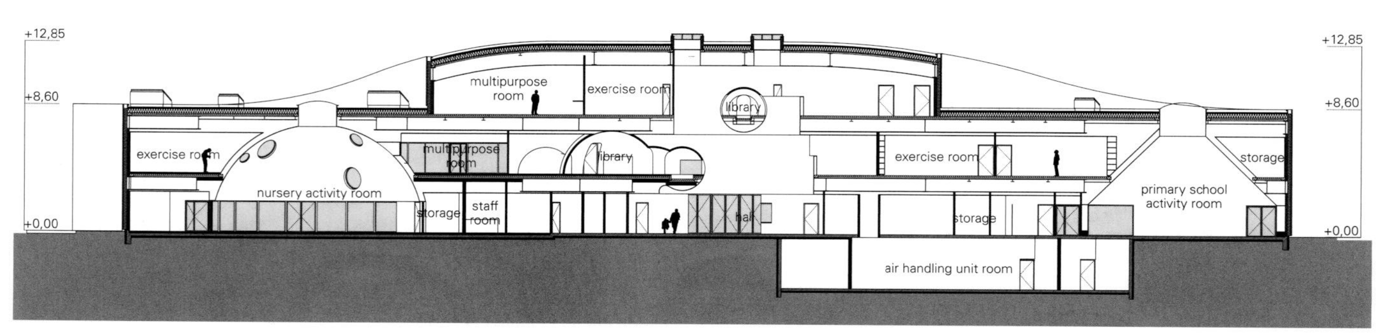

B–B' 剖面图 section B-B'

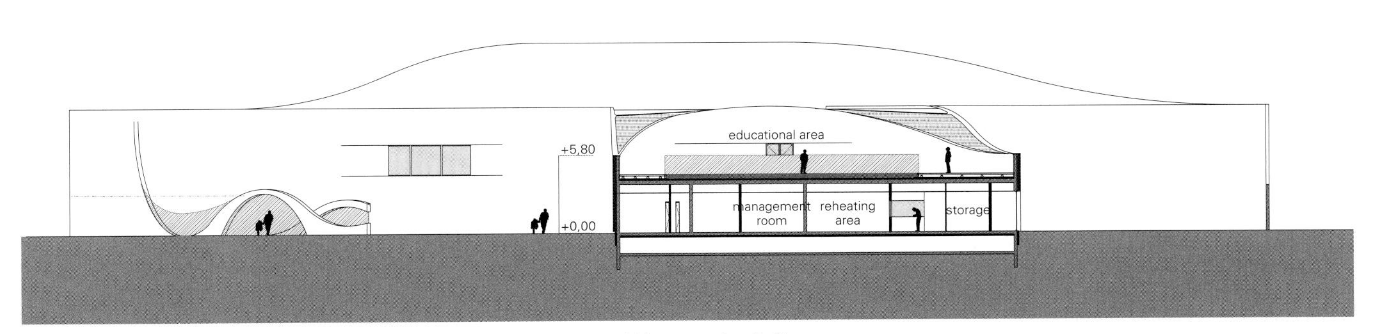

C–C' 剖面图 section C-C'

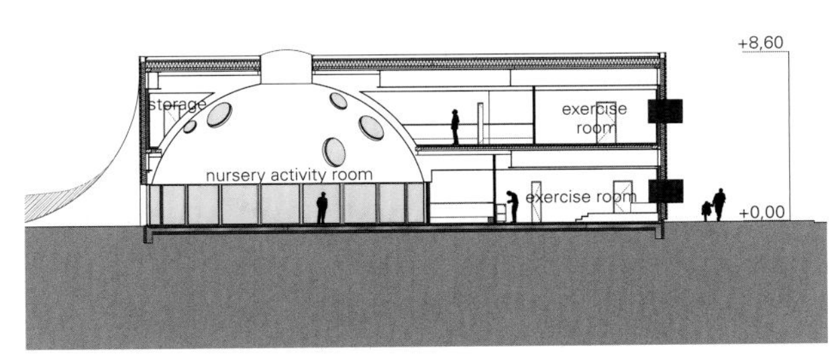

D–D' 剖面图 section D-D'

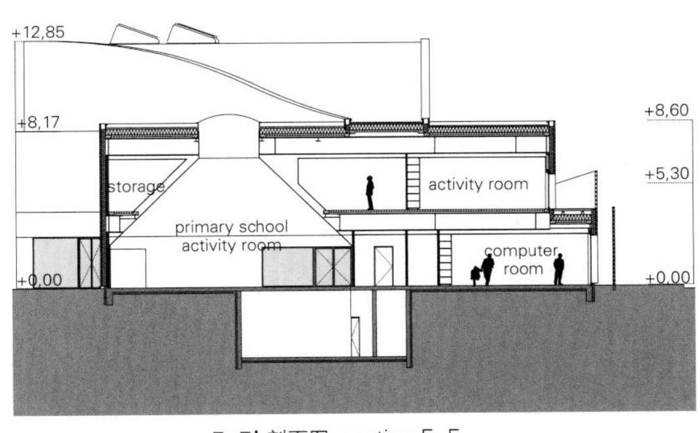

E–E' 剖面图 section E-E

identical, because of their distinct vital functions. This building is symmetrical as well only in appearance.
The nursery classrooms are systematically composed of three circular spaces with three different diameters and ceiling heights. The primary classrooms are square shaped with a side completely "vitrée" (made of glass). The pore spaces between the circles are used for storage allowing more areas for the classrooms.
The multi-purpose nursery room is circular with a half spherical ceiling and the primary one is square with a pyramidal ceiling. These two huge spaces open onto the playgrounds and benefit from large overhead lighting. Also by opening onto the corridors, they contribute to the natural indoor lighting and the ease of orientation in the building with the transparencies between the inside and the outside.
One of our priorities was getting around and ease of orientation in the space. From the entrance, the clover shape structure of the hall enables the visitor to see straight to the two playgrounds. Once in the hall, three accesses converge from the nursery school on the right, the primary school on the left

and the recreation center in the front of the large stairs surrounding the elevator.
The hall pierces the entire height of the building and is lighted by overhead lighting. Two spheres in suspension animate this great void: each sphere is a reading space belonging to the libraries of the nursery school and the primary school.
Due to the clover shape structure, the construction of its facade wasn't easy. To avoid a monotonous and fixed facade, we designed a wooden cladding system with battens that change as the visitor goes around the building: the inferior facet of the battens is painted orange, the superior is painted apple green and the front facets are left rough. This way, the frontal view of the facade is completely neutral and its neutrality recedes as the look becomes oblique. The main facade looks green when we enter the school and orange when we go out. This effect is an initiation to the optic illusions which have great educational value for training the mind. Every choice made on this project has been done considering their impact on children's psycho-motor development. Paul Le Quernec

Haro幼儿园

Taller Basico de Arquitectura

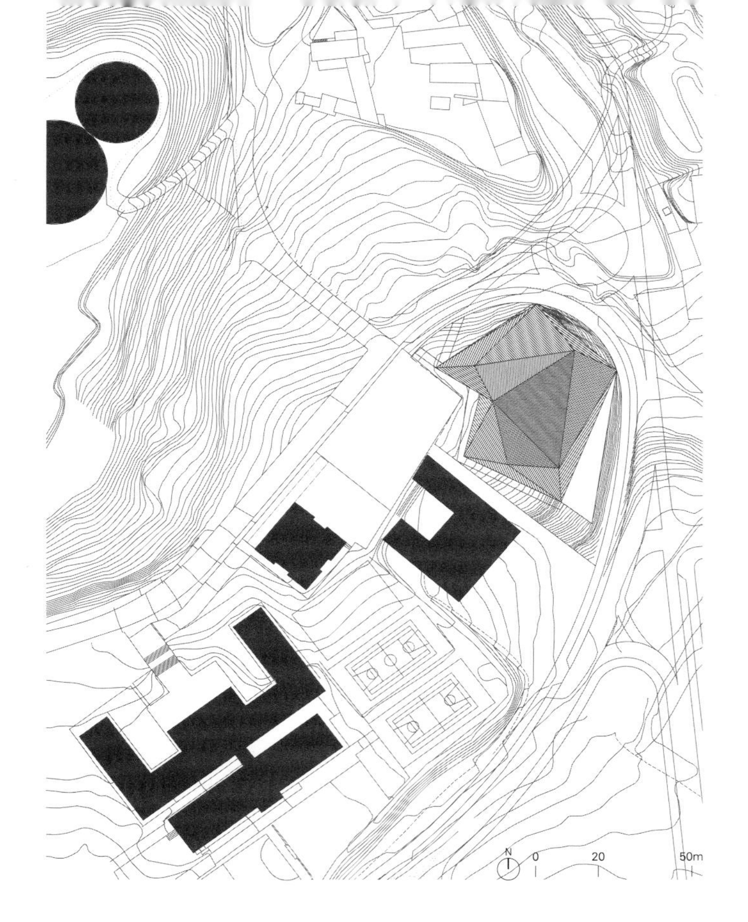

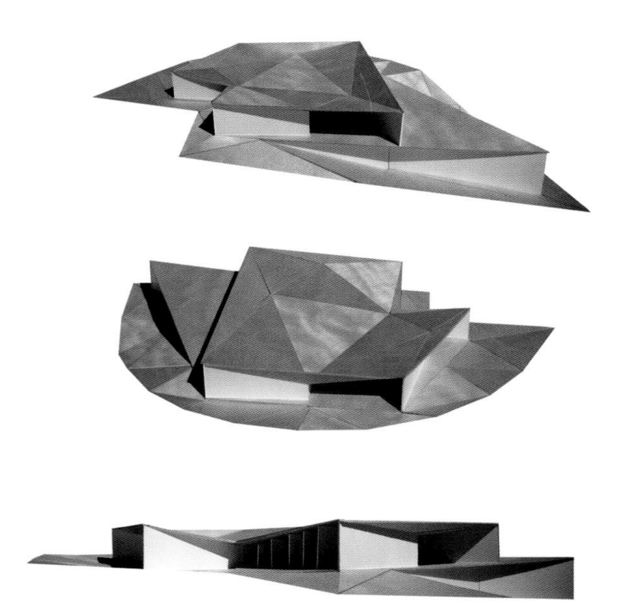

这座幼儿园建在通向山顶的道路之上。其所处的斜坡引导我们来到 Haro 市的边界处。这里没有任何建筑。这是一处奇怪的场地，被道路分割开来。这里没有城市景观，街道和建筑都被空地所取代，被道路围合。这条路似乎是这处新无人城区的标志。场地内仅有的结构便是两个水箱、一个动物收容所以及一座老学校。

这种非常规的城市地形向我们呈现了一个崎岖的山坡，在这里，城市规划师们打算利用仅有的"工具"，即道路系统来对城市进行扩建。他们划定了场地的界限，并将其作为唯一的设计准则。因此，道路将景观改造为一处全新的地形。现在，老城区暂时无法适应新的边界，且这片无人城区也无法容纳典型的城市建筑。

这处全新的城市场地将常规的规划要求与地形法规相结合，因此，典型的城市规划便融入地理学家的地图中，从而形成全新的城市地形。这个全新地形将道路线改造为地形线。而学校，将成为这处新地形的另一个特色。

这座建筑被认为是对矿产资源的一种施工。全新的建筑摒弃了老城的城市风格。因此我们开发了一种新地形建筑，使场地可容纳更多的地形因素。

我们将建筑构思为晶体结构，这种矿物的构架是我们针对这种特殊地形所采取的唯一合适的方案。幼儿园建为一座大型中空岩石结构，是这一区域内非常显眼的矿物结构。

该建筑的设计基础是一组四面混凝土墙体，墙体将教室与办公室分割开来。这些平面将更多的地质情况展示出来。这些地质构造对学校的布局产生了一定的影响，从而促进了全新的内部地质特征的出现。

两面大型水平混凝土立面与场地的陡坡相得益彰。两个楼层与四面呈对角设置的墙体交叉，形成了一系列中空空间。这种结构对内部的地形进行了塑形，并使空间具备城市特色。较低的楼层内设有维修和服务设施，并且和原有斜坡遥相呼应。上层则容纳了学校的教室空间，将这些教室从原有的小山场地中抬高。

Nursery School in Haro

The Nursery School is on the path that leads to the top of a hill. Its ascent takes us to the boundary of the city of Haro. There are no buildings. It is a strange place, fragmented and bisected by a new road. The urban landscape has disappeared, and the streets and the buildings have been replaced by empty spaces, delimited by this road. This road seems to signify a new city that has neither arrived. The only structures are two water tanks, an animal shelter and an old school.
This unconventional urban geography shows us a rugged place of slopes, where urban planners extended the city using as their only tool, the road system. Their hands drew the limits of the plot taken as their only rule. As a result, the road has transformed the landscape into a new geography. And now, the old city does not know how to accommodate this border. This urban no man's land does not allow for the typical architecture of the city.
The new city territory has combined the regular planning laws with topography laws, and as a consequence, the typical city plan has morphed into a geographer's map creating a new urban atlas. This new atlas has transformed the line of the road into a topography line. The school will be another feature of this new atlas.

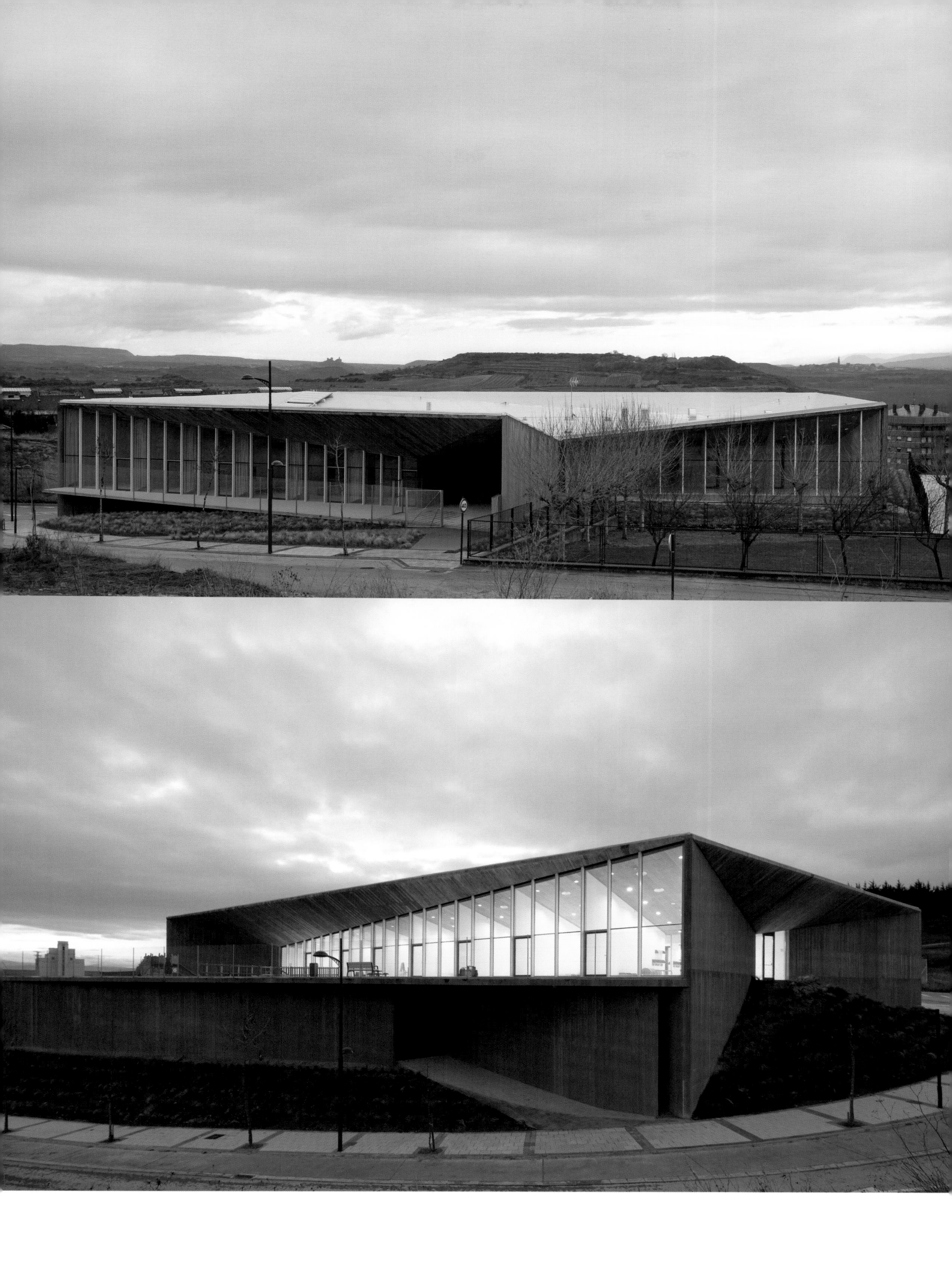

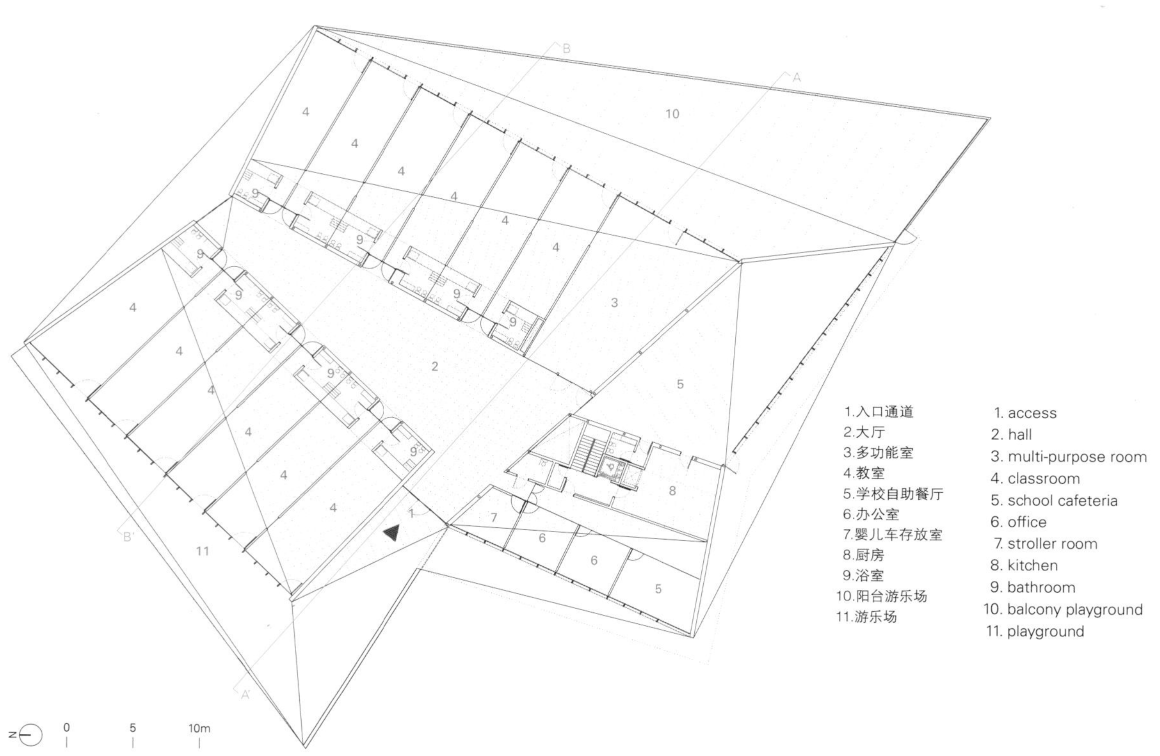
1.入口通道
2.大厅
3.多功能室
4.教室
5.学校自助餐厅
6.办公室
7.婴儿车存放室
8.厨房
9.浴室
10.阳台游乐场
11.游乐场
1. access
2. hall
3. multi-purpose room
4. classroom
5. school cafeteria
6. office
7. stroller room
8. kitchen
9. bathroom
10. balcony playground
11. playground
0
5
10m

The building can be thought of as a construction of mineral source. This new architecture abandons the urban manner of the old city. So we have developed a geographical architecture that considers the location more of a topographical feature.

We have conceived the architecture dictated by crystallographic structures, and this mineral framework leads us to the only coherent solution to the challenge presented by this particular location. The nursery school is now considered as a great hollowed rock which is a visible mineral structure for an area.

The design basis of the building is a set of four concrete walls which divide the classrooms and the offices. These planes reveal more of the mountain's geology. These geological formations influenced the layout of the school, prompting new internal geological features.

Two great horizontal concrete facades compliment the steep slopes of the site. The two levels intersect with the four diagonal walls providing a sequence of hollow places. This structure gives shape to the internal geography, and opens the spaces to the urban topography. In the lower level, housing maintenance and service facilities are anchored by the topography of the original slopes. The upper level accommodates the school classrooms, elevating them from the existing hill.

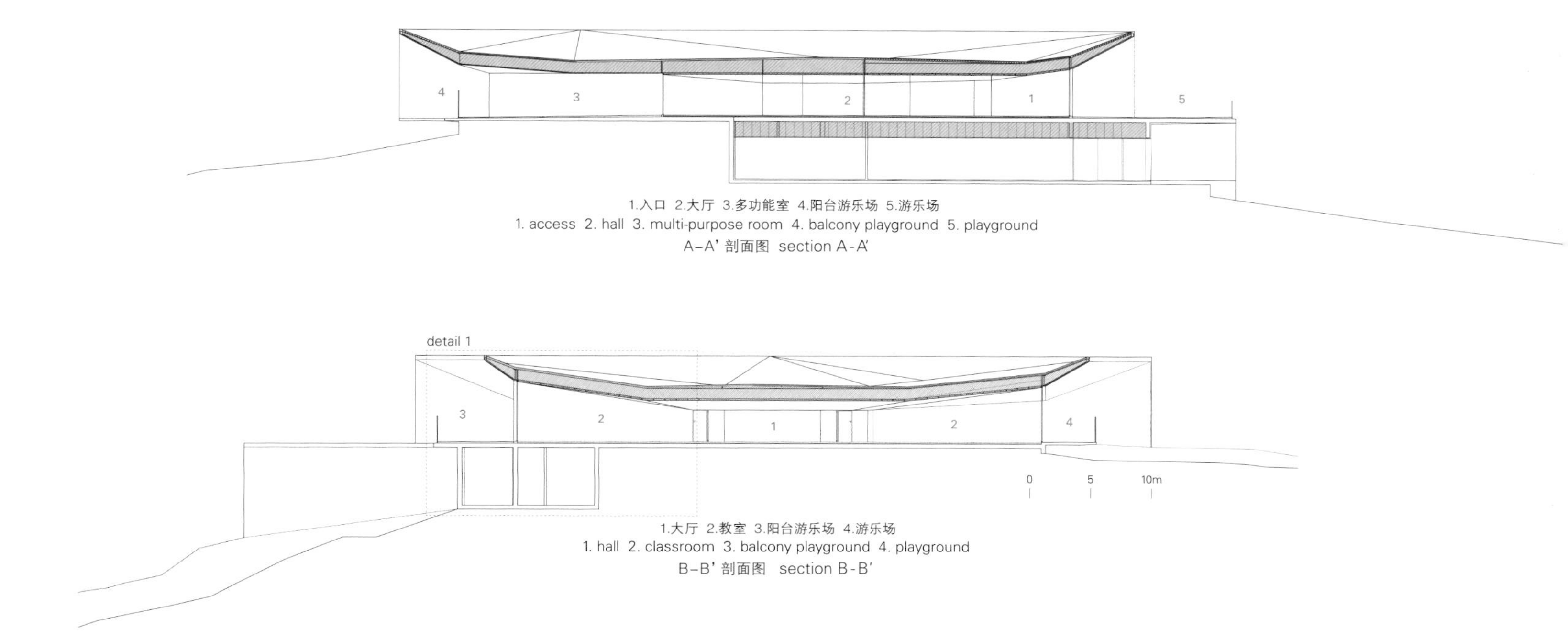

1.入口 2.大厅 3.多功能室 4.阳台游乐场 5.游乐场
1. access 2. hall 3. multi-purpose room 4. balcony playground 5. playground
A-A' 剖面图 section A-A'

1.大厅 2.教室 3.阳台游乐场 4.游乐场
1. hall 2. classroom 3. balcony playground 4. playground
B-B' 剖面图 section B-B'

项目名称：Nursery School in Haro, Spain / 地点：C/ Gonzalo de Berceo, Haro, La Rioja, Spain
建筑师：Taller Básico de Arquitectura / 首席建筑师：Javier Pérez Herreras, Javier Quintana de Uña
合作：Edurne Pérez Díaz de Arcaya, Manuel Antón Martínez, Xabier Ilundain Madurga, Joseba Aramburu Barrenetxea, David Santamaria Ozcoidi, Laura Elvira Tejedor
结构工程：FS Estructuras /设备工程：GE & Asociados / 建筑工程：Carlos Munilla Orera
客户：Regional Ministry of Education, Culture and Tourism of the Government of La Rioja
施工单位：DAYCO Rioja, LMB group / 用途：School(0~3 years)
用地面积：3,608m² / 总建筑面积：2,490m² / 造价：EUR 2,291,233.74 / 施工时间：2011.10—2013.12
摄影师：©Pedro Pegenaute (courtesy of the architect)

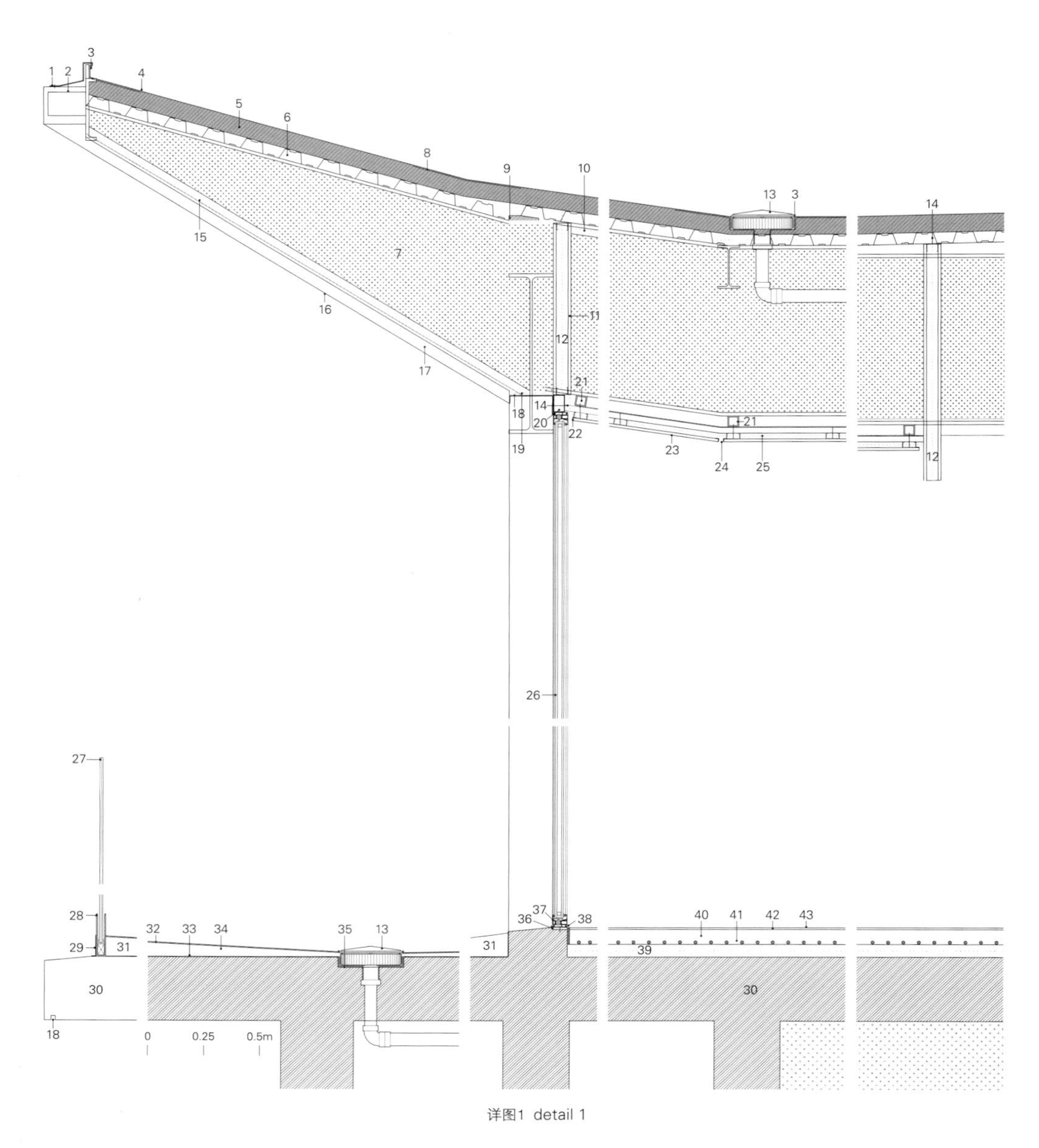

详图1 detail 1

1. coping in thermo lacquered galvanized steel folded sheet metal 3mm thk. ral 9006
2. reinforced concrete beam
3. seal
4. PVC waterproof membrane
5. EPS thermal insulation 40+40mm thk.
6. steel folding steel sheet
7. laminated steel beam
8. waterproof reinforcement
9. metallic wedge
10. extruded polystyrene foam thermal insulation 5cm thk.
11. plasterboard 13+13mm fixed to galvanized steel hidden structure
12. rock wool blanket thermal insulation 5cm thk.
13. geberit drain
14. projected polyurethane
15. reinforced concrete. 10cm thk.
16. colorless acrylic application
17. reinforced concrete filler
18. groove
19. insulation
20. steel tube 80.50.2.
21. steel tube 50.50.5 every 80cm
22. aluminium L profile ral 9010 2mm thk.
23. suspended plasterboard ceiling with acoustic properties
24. open joint
25. rock wool blanket thermal insulation 3cm thk.
26. double glazing (8+8)/15/(6+6)
27. laminated glazing (5+5)
28. galvanized steel L profile 100.50.6.
29. galvanized steel L profile 20.20.3.
30. floor mortar bed for slope
31. child paving 5mm thk.
32. rubber waterproofing layer 1,2mm thk.
33. geotextile layer, 200g/m² density
34. waterproof reinforcement
35. steel galvanized plate 10mm thk.
36. aluminium windows frame with thermal break TECHNAL FXI65.
37. expansion gap
38. high density EPS 25kg/m³. 6cm thk.
39. aligning mortar 7cm thk.
40. underfloor heating system
41. self-levelling mortar. 1cm thk.
42. continuous floor in white PVC with
43. acoustic properties

普兰京斯幼儿园

Pierre-Alain Dupraz

© Fernando Guerra

新普兰京斯沙幼儿园的设计构思是在倾斜场地内建造一座大房子。该结构的体量相互交叉，因此呈十字形，与周围整体环境形成了一种特殊的关系。

建筑的四个体量交错衔接，每部分之间存在三分之一单元层的高度差，这种做法使建筑整体看起来更小巧，视觉上也感觉不突兀。

六间教室两两一组，分别布置在三个底层体量中，楼梯则作为幼儿园的基础结构，位于建筑的中央位置。与楼梯平行设置的斜坡也提供了进入建筑的通道，同时这条坡道也是通向幼儿园的内部入口。

婴儿室所在的体量与幼儿园体量垂直相交，强调并遮蔽了入口区域。人们可直接通往建筑的上层，因此婴儿室这一体量可以独立使用，这一层由大型房间（餐厅）构成，这些大房间可以根据不同的需要灵活分割成若干个小单元。

朝向不同方向的大型采光窗更加凸显了场地的自然特色，因为建筑师非常注重建筑与场地的视觉关系。

建筑的室内和室外均采用了混凝土材料，使之与地形融为一体。

Prangins Kindergarten

The new Prangins kindergarten is conceived as one big house that has been placed on a sloping site. The cruciform structure resulting from the interlocking volumes allows it to have a special relationship with the neighbouring ensemble.
The four volumes, which are mutually staggered by a third of each unit's floor height and interleaved, make the building appear to be smaller and visually less obtrusive.

N 0 20 50m

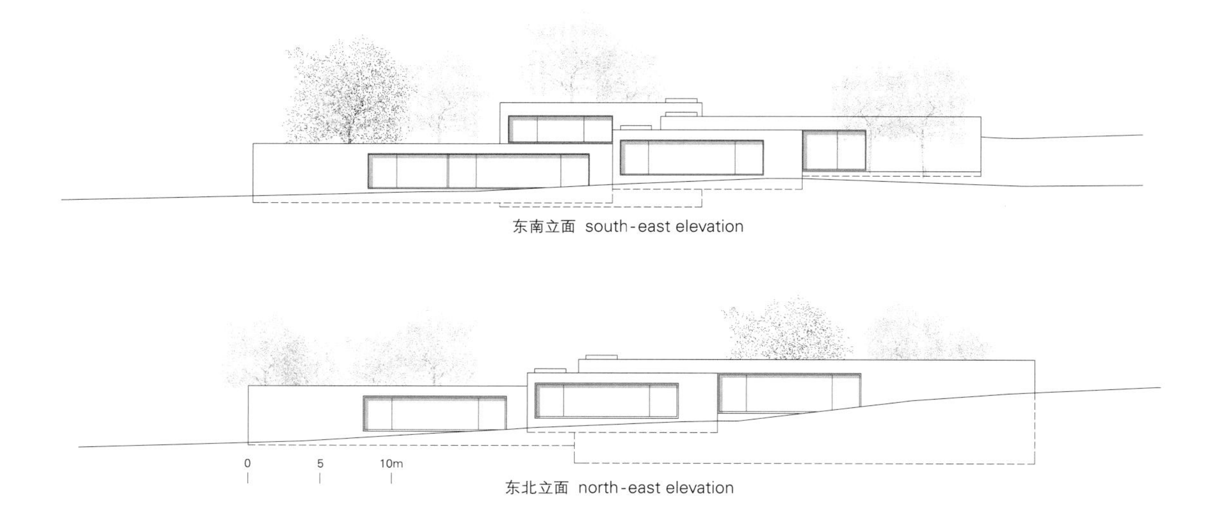

东南立面 south-east elevation

东北立面 north-east elevation

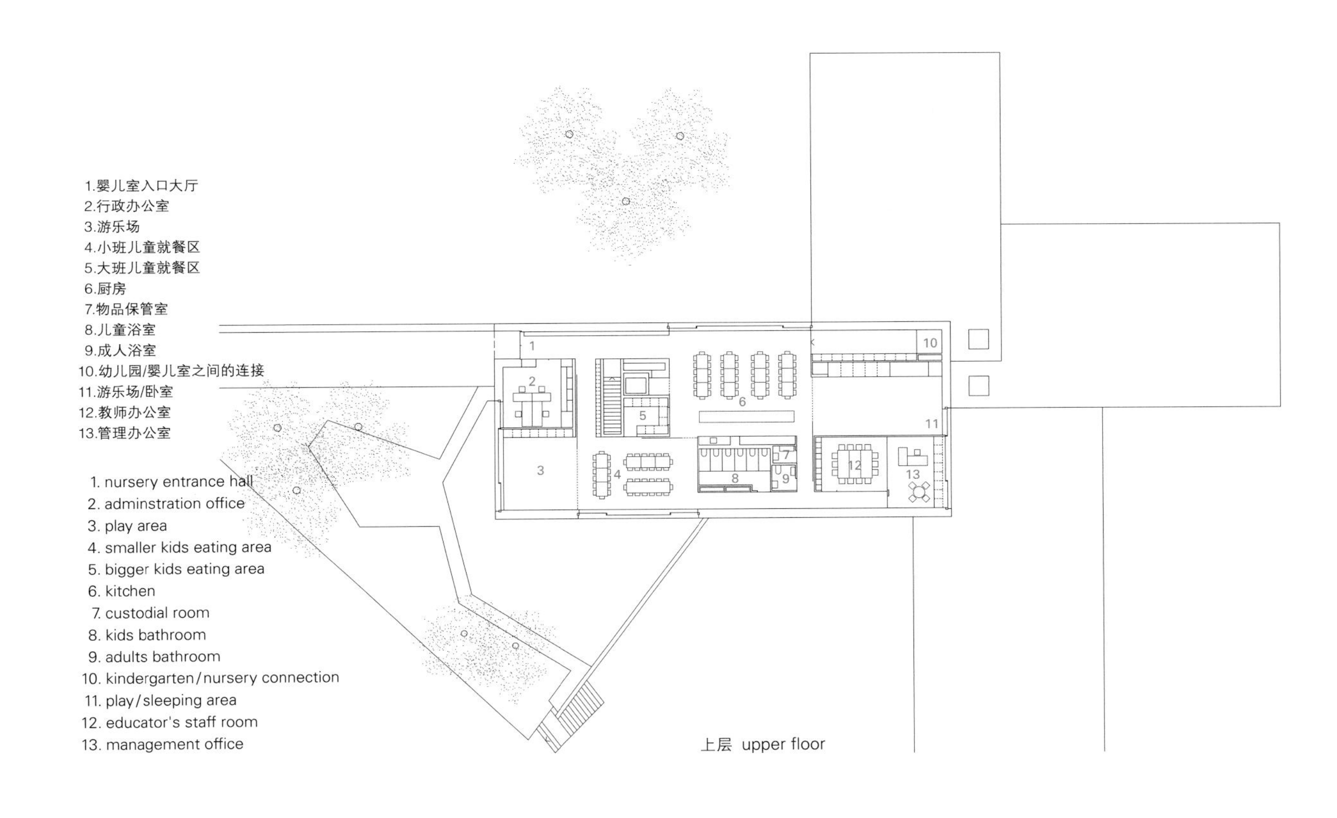

上层 upper floor

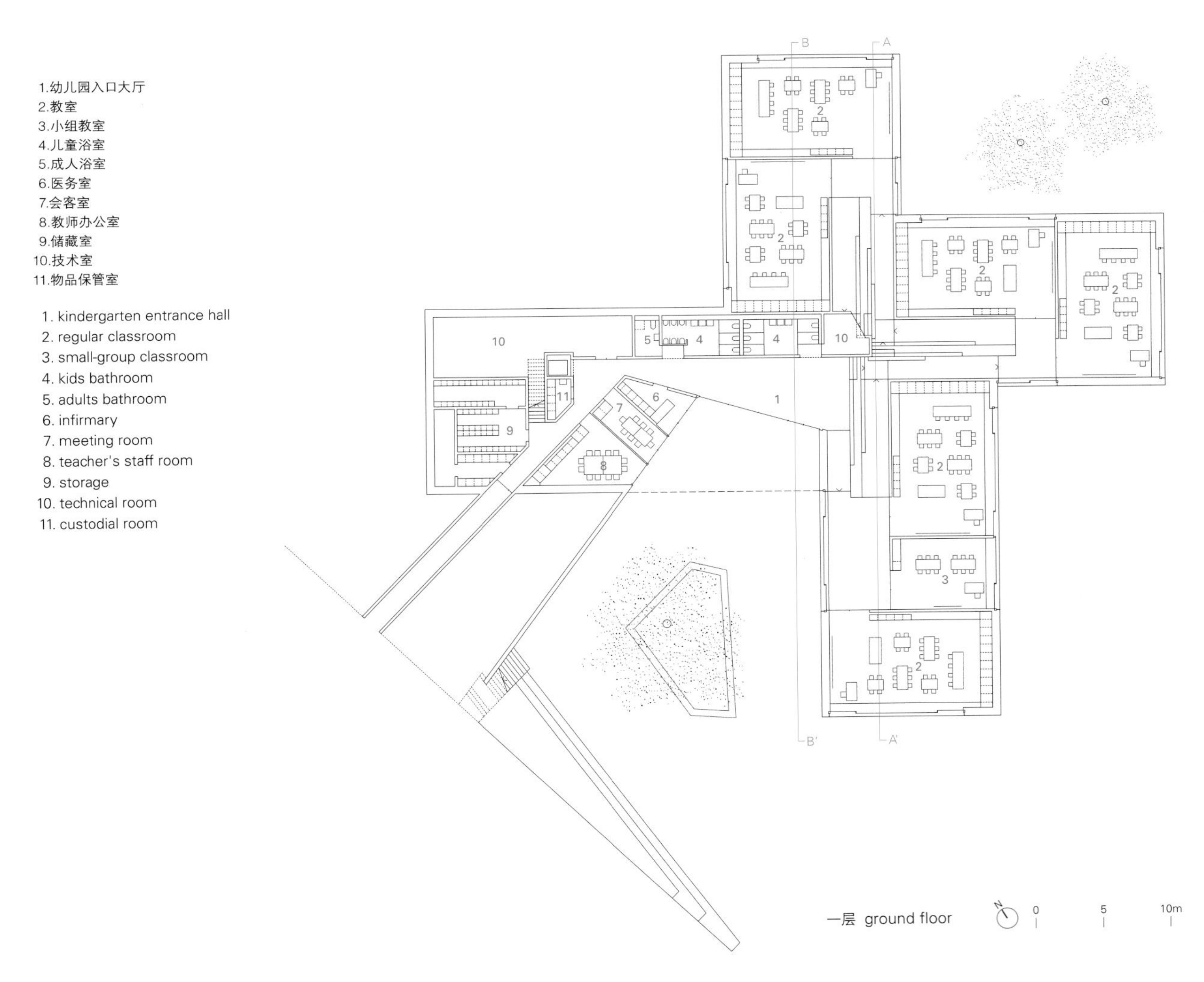

一层 ground floor

项目名称：Kindergarten and Crèche / 地点：Prangins (Vaud), Switzerland / 事务所：Pierre - Alain Dupraz / 项目建筑师：Nicola Chong, Frederico Vieira / 项目团队：Julian Behrens, Maxime Beljansky, Kira Graf, Paolo Marchiori, Pierre Mencacci / 用地面积：7,218m² / 总建筑面积：1,248m² / 有效楼层面积：1,692m² / 竞标时间：2011 / 施工时间：2013~2015 / 摄影师：©Thomas Jantscher (courtesy of the architect) (except as noted)

The six classrooms are organised in groups of two and occupy the three lower volumes. Due to its central location, the staircase is a fundamental, structuring element of the kindergarten. The ramps that accompany it provide a parallel access and act as an internal entrance to the crèche.
The volume of the nursery, which is placed perpendicular to the first wing, marks and covers the entrance area of the kindergarten. Direct access to the upper part of the property also enables independent operation of the nursery. The nursery floor plan consists of a large (dining) room, which can be divided into several smaller units according to requirements. The natural qualities of the location are highlighted by the different alignments of the large window openings, since the design paid great attention to visual relationships with respect to their placement.
The reinforced concrete that is visible both inside and outside, anchors the building into the topography.

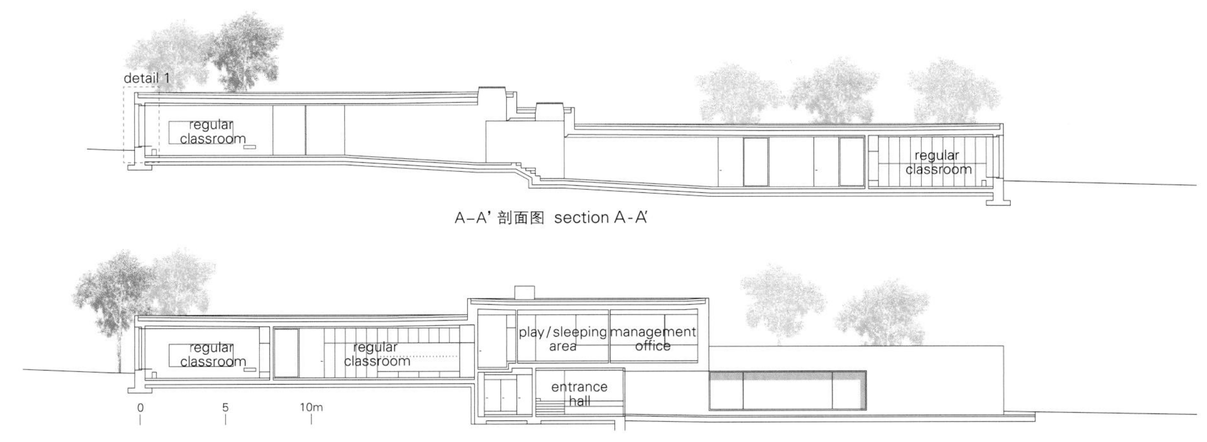

A-A' 剖面图 section A-A'

B-B' 剖面图 section B-B'

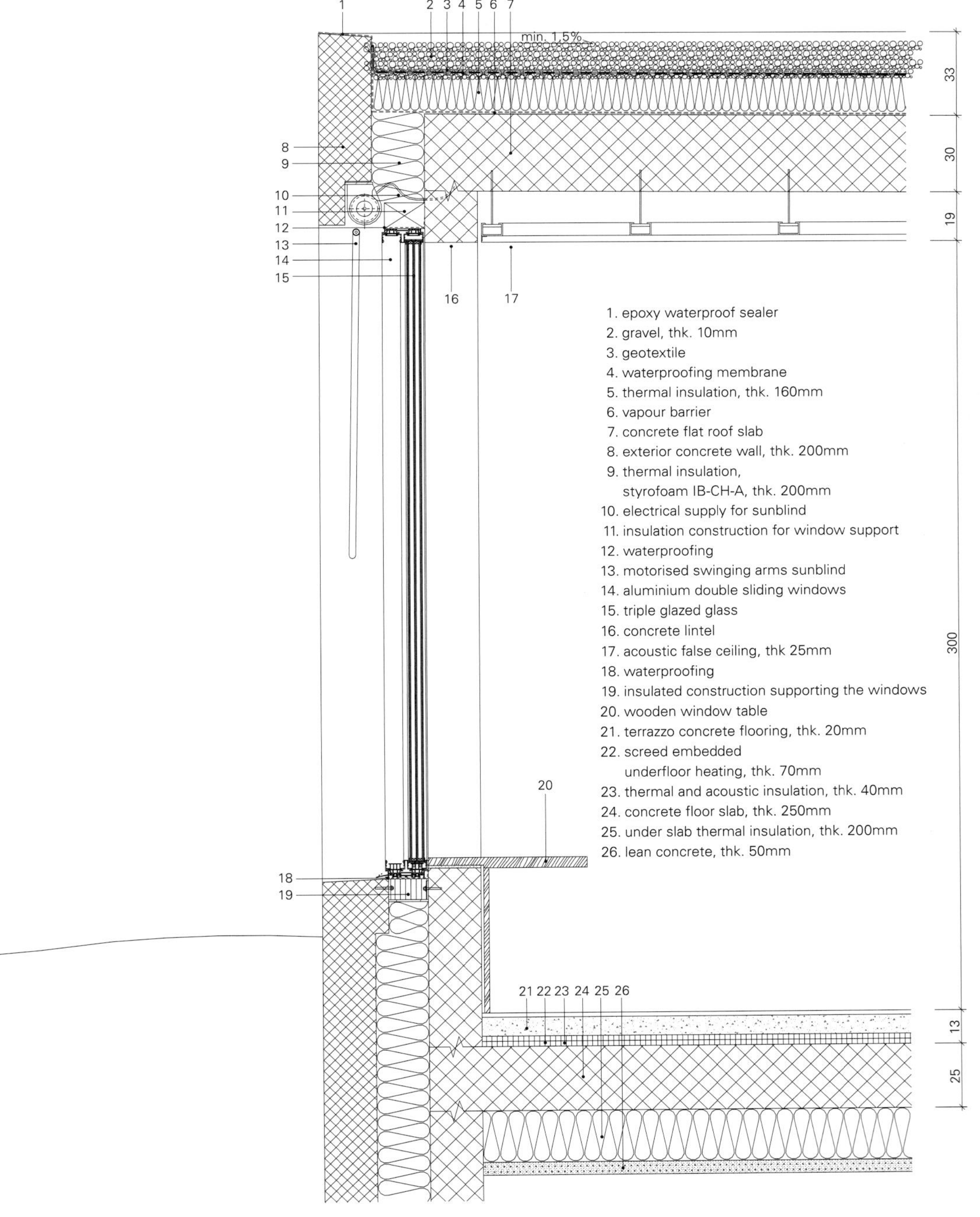

1. epoxy waterproof sealer
2. gravel, thk. 10mm
3. geotextile
4. waterproofing membrane
5. thermal insulation, thk. 160mm
6. vapour barrier
7. concrete flat roof slab
8. exterior concrete wall, thk. 200mm
9. thermal insulation, styrofoam IB-CH-A, thk. 200mm
10. electrical supply for sunblind
11. insulation construction for window support
12. waterproofing
13. motorised swinging arms sunblind
14. aluminium double sliding windows
15. triple glazed glass
16. concrete lintel
17. acoustic false ceiling, thk 25mm
18. waterproofing
19. insulated construction supporting the windows
20. wooden window table
21. terrazzo concrete flooring, thk. 20mm
22. screed embedded underfloor heating, thk. 70mm
23. thermal and acoustic insulation, thk. 40mm
24. concrete floor slab, thk. 250mm
25. under slab thermal insulation, thk. 200mm
26. lean concrete, thk. 50mm

详图1 detail 1

Amanenomori幼儿园

Aisaka Architects' Atelier

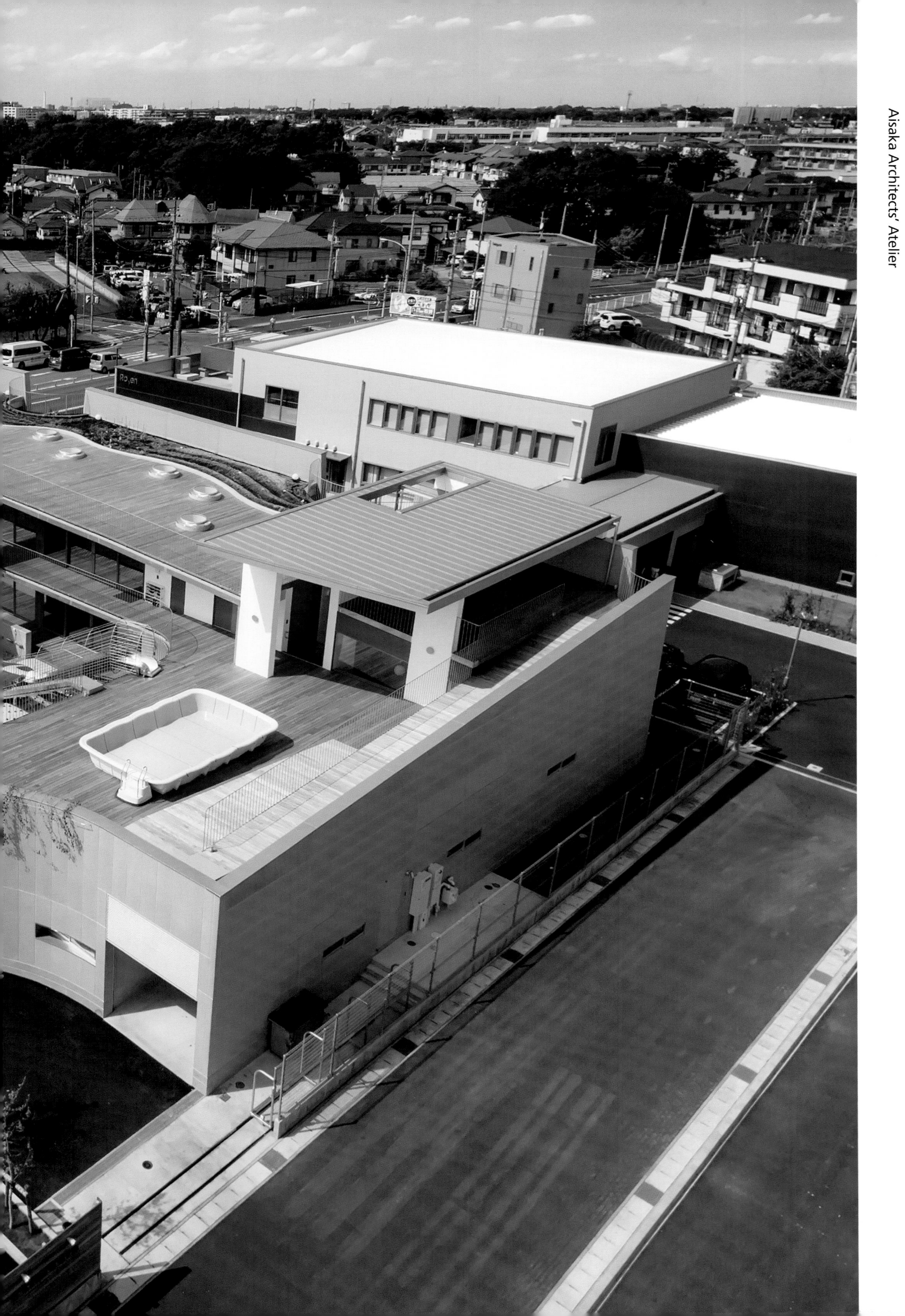

"弥（Amane）"是一个意思为"圆形""环绕""周围"的日本汉字，代表着这个幼儿园对孩子的祝福。希望孩子们能感受到来自围绕他们的自然环境的庇佑和树林周围的环形建筑特色。我们希望孩子们能在建筑物的内部和外部散步，感受周围的一切，并培养他们的感受力和思考能力。

这个双层幼儿园坐落在船桥市，其特点是带有立体式环形结构，以及一个屋顶露台。我们的设计理念是为 160 名儿童提供足够的自然玩耍空间，并让所有的家长和幼儿园工作人员感到安心。

场地南侧的四分之一区域用于设置入口人行道，其余的部分作为幼儿空间。办公人员、幼儿园老师和厨师们的办公室位于入口和幼儿空间之间的边界，以实现简洁性和安全性。我们设计的环形结构不仅为孩子们提供了充满趣味的游乐场，而且还设有发生危险时快速通向紧急疏散路线的通道。建筑中央为庭院，外围种植着一些树，并沿着环形结构在二者之间设置了露天平台，斜坡、楼梯和桥。整体结构被梯形实体墙体和外部屋顶所覆盖，该结构保护了孩子的安全玩耍。这个围绕庭院建造的圆形建筑设有带防雨屋檐的走廊，为成人提供舒适感和安全感。从另一方面，整个结构的设计便于忙碌的父母在幼儿园工作人员的帮助下快速接送他们的孩子，免去了脱鞋进入的麻烦。

每层的外部空间不仅是单纯的开放空间，更有诸多的变化。如丰富多样的阳光和阴影区、高起的屋檐和一些屋檐下狭窄的空间、斜坡、山丘以及通过改变楼层和屋顶的高度和方向来形成的空腔。孩子在这样的环境下一年四季都不会感到无聊。半圆形的花园中心的绿植不仅能为整座建筑提供良好的通风，还可以增进孩子们与自然的感情。为了培养孩子们对食物的感情和兴趣，我们在屋顶建造了一个蔬菜园，在一层安装一间透明的玻璃墙厨房，为孩子提供饮食方面的教育。厨房楼层平面略低于外侧，便于孩子们观察内部，同时也方便工作人员留意在庭院玩耍的孩子们，消除办公室视角的盲点。

墙体和栏杆的圆形倒角设计主要是出于对安全的重视，与此同时，建筑所有天窗的边角也是如此。

为了节能，我们调整了屋檐来控制光照，花园可改善通风条件，屋顶露台和蔬菜园能为屋顶提供保温功能，地热管道系统可提供地热能，河和小水池可用于循环利用雨水，太阳能电池板则可生产再生能源。通过观察这些构件的日常运用，孩子们可以更好地了解"自然"，包括有关植物和风雨的现象。

为了让孩子们有机会去通过感受材料的质感来学习不同材料的名字，我们尝试让"木像木头，钢铁像钢铁，石头像石头"，来保持每种材料的原始质感。从这个角度来说，我们不用原色，相反地，我们建造了具有高度对比性的结构，利用其多样化的空间特色和环境来获得鲜明对比的体验。

Amanenomori Nursery School

"Amane" is one of the Japanese kanji that stands for "round," "around" or "all-around". It represents the wish of the nursery school to let children feel the blessing of the all-around nature and also its architectural feature of circular shape around the woods. We wish that children can go around both inside and outside of the building, feeling everything around, and nourish their sensibility and ability to think.

A nursery school of two-story building with rooftop terrace features 3-dimensional and circuit style structure located in Funabashi city. The concept of our design is to provide enough space for 160 children to play around the nature and for all their parents and nursery staff to feel safe.

The south quarter of the site is used for entrance walkway, and the rest of the part is for nursery space. Several rooms for office staff, nursery staff and cooks on the border between entrance and nursery space achieve both simplicity and security. We designed the circular ring-shaped structure that provides enjoyable playground for children and easy access to escape route in case of emergency, having the courtyard in the middle, planting trees along the outer edge, and installing the deck, slopes, stairs, and the bridge along the circle between them. Covered with the solid trapezoid-shape wall and

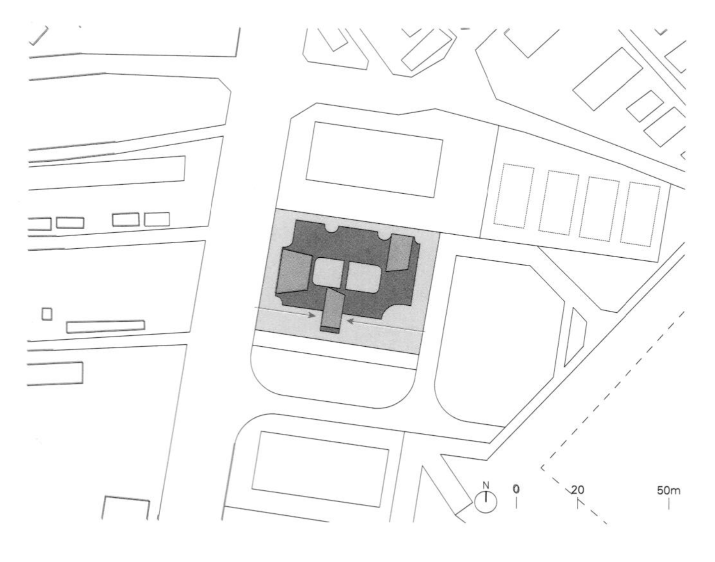

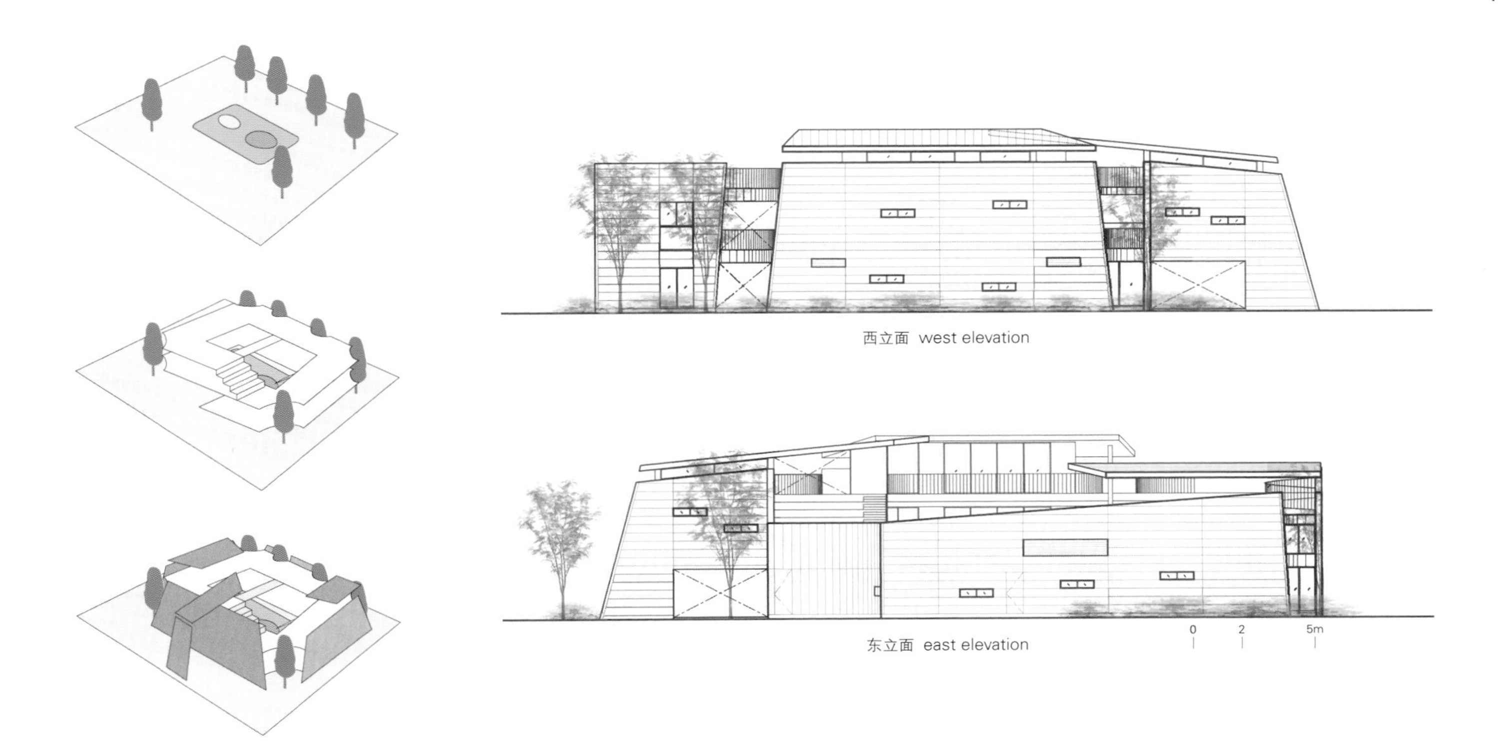

西立面 west elevation

东立面 east elevation

roof outside, its overall structure achieves to protect children's pleasure with its strength. Its O-shaped building surrounding the courtyard with outside corridor with eaves for weather protection also provides comfort and a sense of safety to adults. This structure helps busy parents to drop off and pick up their children quickly without taking off shoes and nursery staff to help each other on the other side.

Outside space of each floor not just opens to outside, but also provides various changes, such as sunny spot and shade, higher eaves and narrower space under eaves, slopes, hills and cavities produced by changing the direction and the height of floors and roofs, so that children spending the whole year here do not get bored. The half-circle-shaped garden to help ventilation also delivers affection toward nature with green plants in the center. From the perspective of dietary education to develop children's appreciation and interest toward food, we place the vegetable garden on the rooftop and a glass-walled kitchen on the first floor. The floor level of kitchen is settled lower to let children look into kitchen, at the same time, it is able to keep an eye on the courtyard to compensate for blind spot from the office.

Round chamfers of walls and railings are necessary for safety reasons and also for the edge of light and skylight in every part of the building.

For an incorporated energy saving scheme, we adopt the eaves to control sunlight, the garden to improve ventilation, the rooftop deck and vegetable garden for heat insulating of rooftop, the earth tube heating system to use geothermal heat, the river and the pond to reuse the rainwater, and solar panels to produce circulating power. Watching these structures in daily life, children can learn about "the nature" including phenomenon about plants or the wind and rain.

In order to give children the opportunity to learn the name of materials to feel its original texture at the same time, we try to use "wood as wood-like, steel as steel-like and stone as stone-like" to keep the original texture of each material. From this perspective, we didn't use the primary colors. Instead, we exploit the high contrasted structure to provide contrasting experience using the various spatial features and environments. Aisaka Architects' Atelier

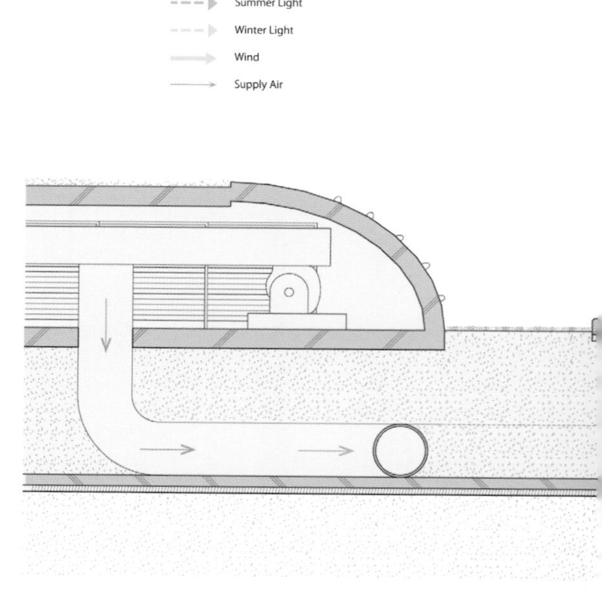

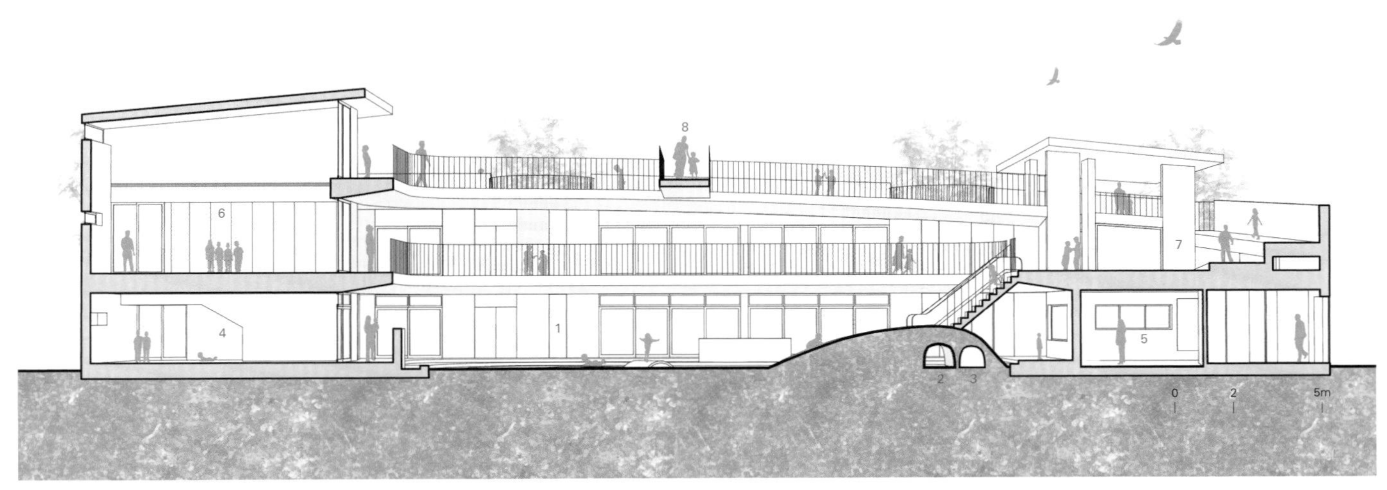

1. 庭院 2. 隧道 3. 机械室 4. 婴儿室 5. 厨房 6. 游戏室 7. 露台 8. 桥
1. courtyard 2. tunnel 3. machine room 4. baby nursery room 5. kitchen 6. playroom 7. sunny terrace 8. bridge
A-A' 剖面图 section A-A'

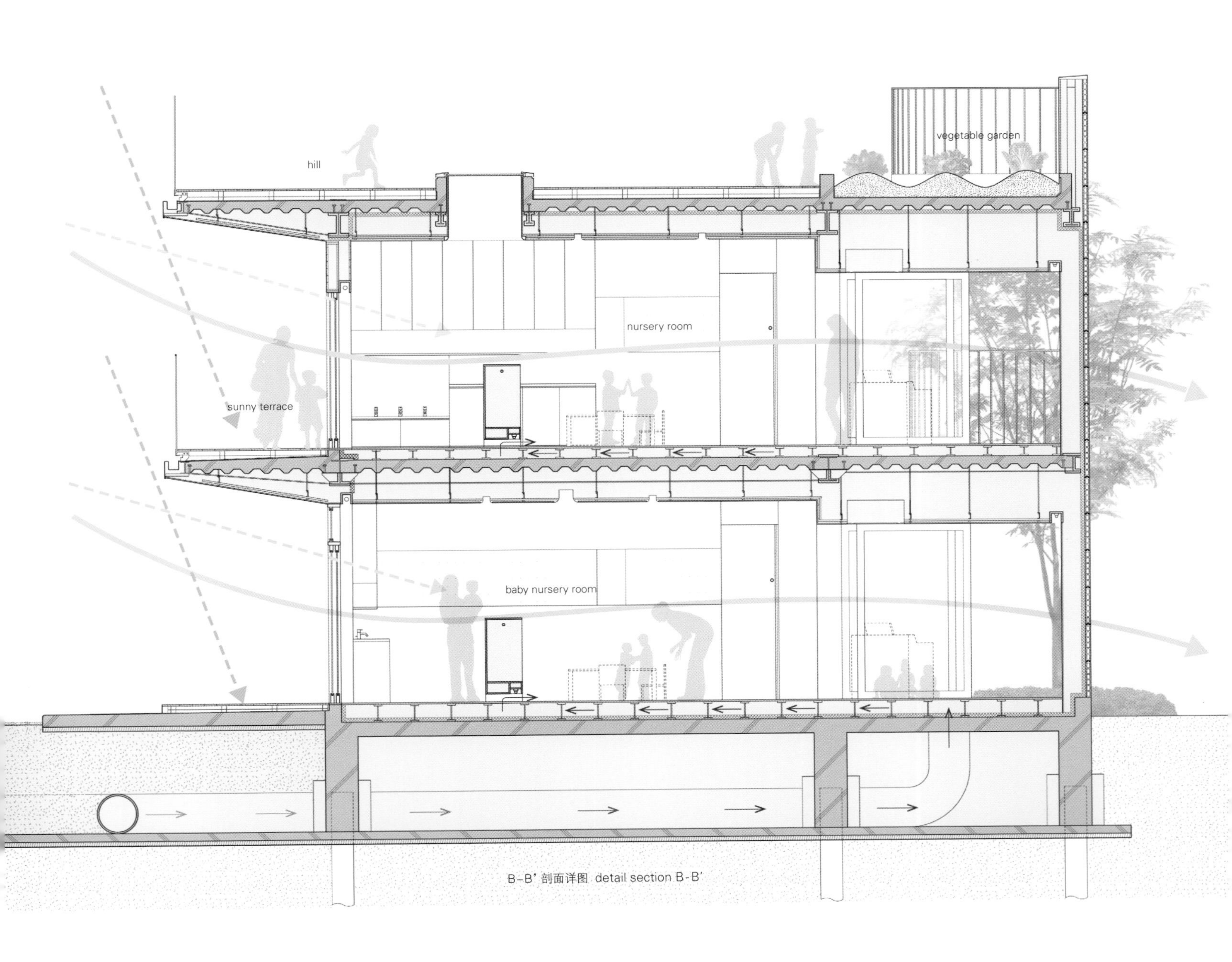

B-B' 剖面详图 detail section B-B'

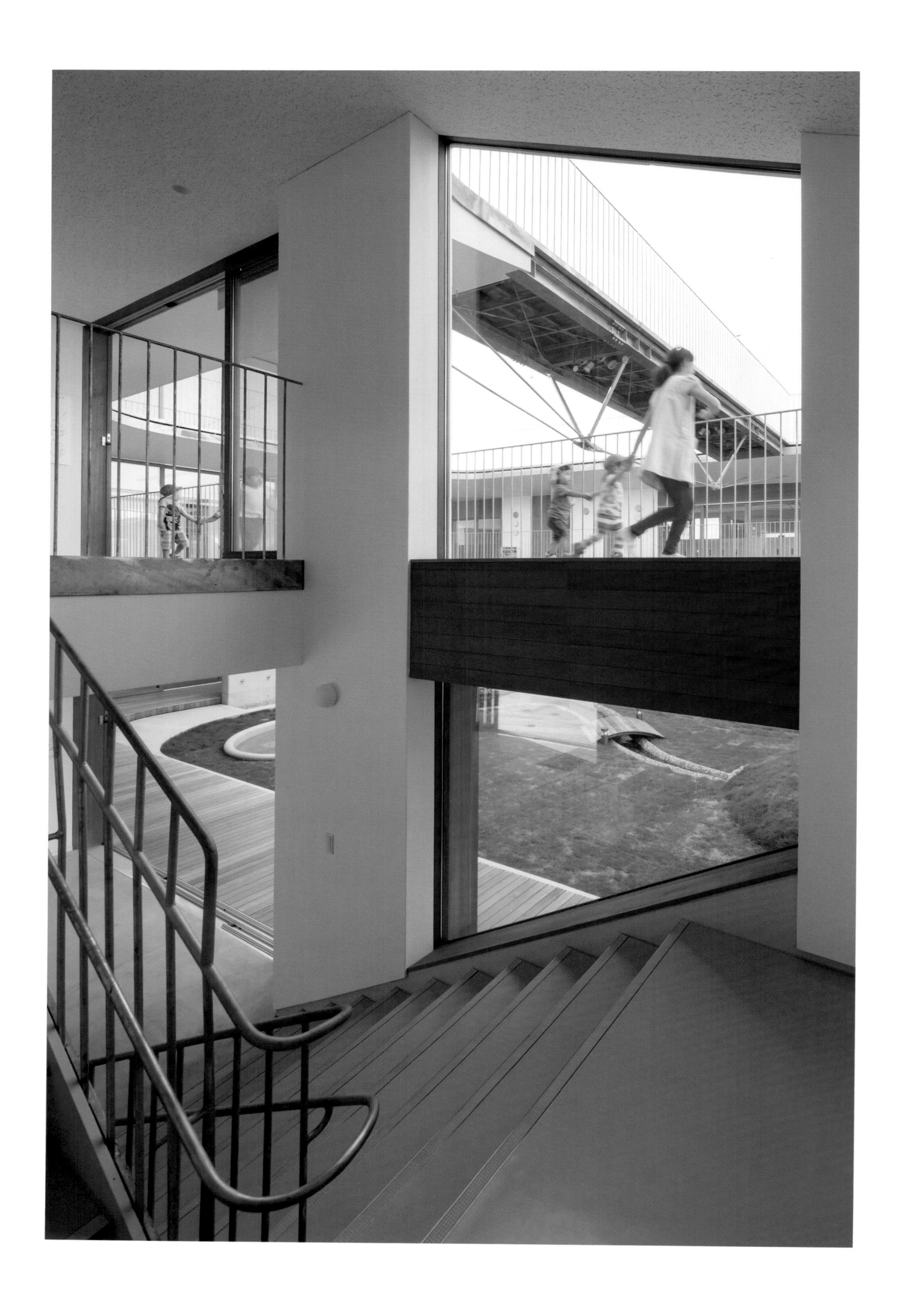

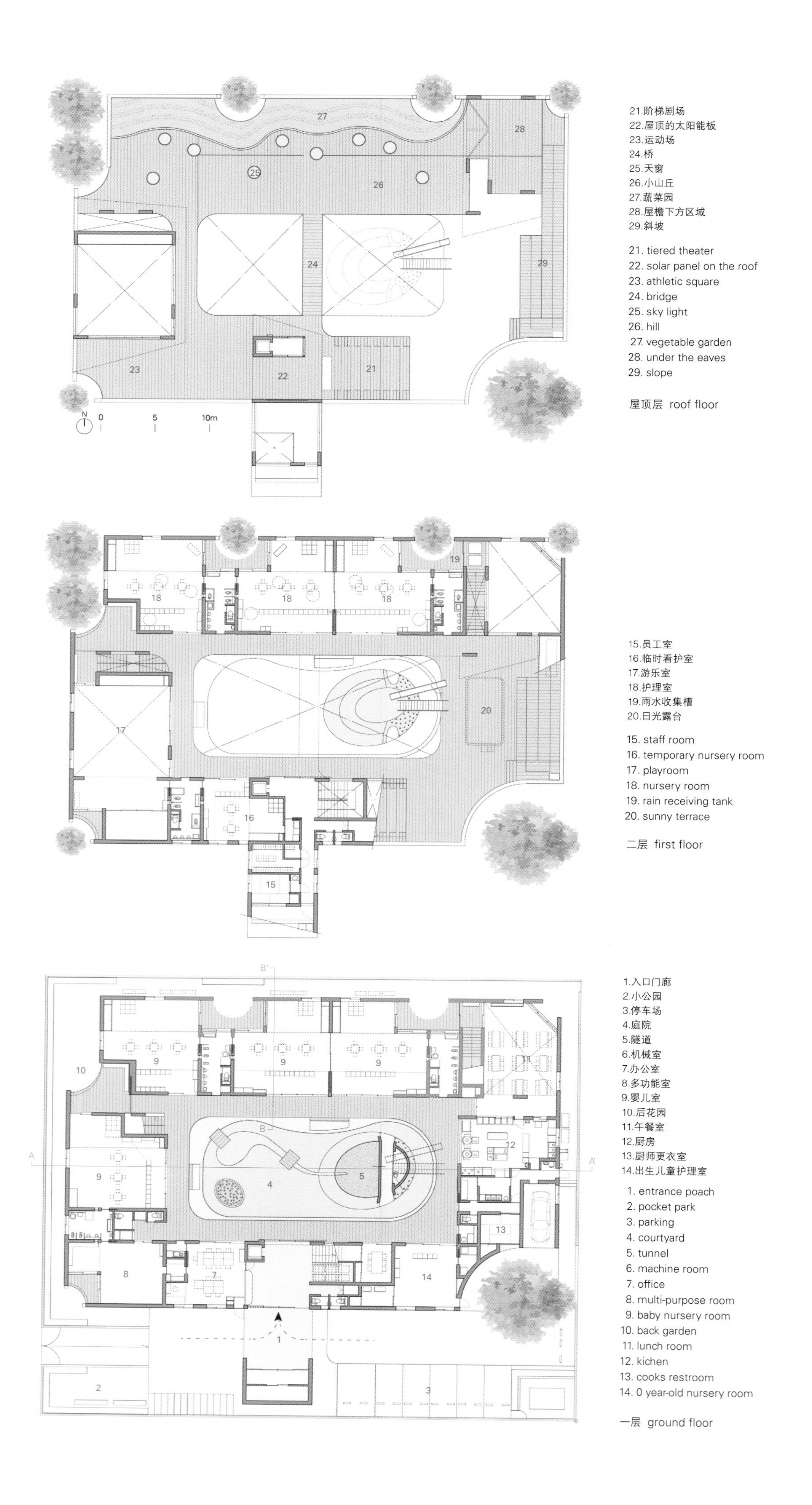
21.阶梯剧场
22.屋顶的太阳能板
23.运动场
24.桥
25.天窗
26.小山丘
27.蔬菜园
28.屋檐下方区域
29.斜坡
21. tiered theater
22. solar panel on the roof
23. athletic square
24. bridge
25. sky light
26. hill
27. vegetable garden
28. under the eaves
29. slope
屋顶层 roof floor
N
0
5
10m
15.员工室
16.临时看护室
17.游乐室
18.护理室
19.雨水收集槽
20.日光露台
15. staff room
16. temporary nursery room
17. playroom
18. nursery room
19. rain receiving tank
20. sunny terrace
二层 first floor
1.入口门廊
2.小公园
3.停车场
4.庭院
5.隧道
6.机械室
7.办公室
8.多功能室
9.婴儿室
10.后花园
11.午餐室
12.厨房
13.厨师更衣室
14.出生儿童护理室
1. entrance poach
2. pocket park
3. parking
4. courtyard
5. tunnel
6. machine room
7. office
8. multi-purpose room
9. baby nursery room
10. back garden
11. lunch room
12. kichen
13. cooks restroom
14. 0 year-old nursery room
一层 ground floor

项目名称：Amanenomori Nursery School / 地点：Funabashi, Chiba, Japan / 建筑师：Aisaka Architects' Atelier_Kensuke Aisaka / 结构工程师：Kanebako Structural Engineers /用途：Nursery school / 用地面积：2,051.59m² / 总建筑面积：1,067.41m² / 有效楼层面积：1,493.54m² (824.52m²_1F/624.38m²_2F, 44.64m²_PH) / 结构：steel / 室外饰面：galvanized steel plate, wood deck_roof/extruded cement panel_wall / 室内饰面：birch flooring_floor/plasterboard, wood siding_wall / 竣工时间：2015.7 / 摄影师：©Shigeo Ogawa (courtesy of the architect)

儿童日托中心

Burobill & ZAmpone architectuur

这座儿童日托中心位于布鲁塞尔的中心，可容纳 68 名儿童。虽然街区场址的空间有限，但是它仍然包括三座学校、一座音乐学院，课后托管中心和多个机构。这个日托中心将会全权负责从孩子们进入托管中心到他们满 18 岁离开托管中心的所有教育任务。

该项目坐落在一个现有建筑街区的一楼。孩子和家长可以通过一条水泥路进入托管中心这一层。托管中心二层的露台也是水泥建造，且富有曲线形，它是 68 名儿童的户外操场。

该建筑具有一个开放的楼层平面，儿童的卧室都集中设置。房间由滑动门连接，使这些房间可以根据孩子们的需要进行结合、分组和分隔。建筑的南侧设有一条带大型窗户的走廊。这是一处可获得热量的缓冲区，同时也是一处孩子骑车的游乐区，我们将这条走廊称之为"游乐街"。这条"游乐街"利用大型窗户在一年四季都能获得足够的阳光和热量，因此，它是一处非常完美的游乐区。

沿着南侧立面设置的一个小型木凳可用于孩子们休息和玩耍。这个长凳设计使孩子们在玩耍累了之后或者等待的时候可以坐下来休息。

为了在现有立面中切割出大型洞口，来为孩子们的游乐区留出空间，游乐区内需要安装一些柱子。这些柱子都采用镜面抛光处理，以便和孩子的玩耍融为一体：孩子们可以在柱子表面看见自己。

斜坡和户外操场的栏杆经过精心设计，全部采用水泥制成，来给予孩子们安全感。而成人在此站立时，也能俯瞰外面的世界。

Child Day Care Center

The Child Day Care Center will house 68 children in the center of Brussels. There is limited space available in this building block that also hosts 3 schools, a music academy,a after-school care & many more organizations to guide children from the day they enter into Day Care till the day they leave school at the age of 18.

This project is situated on the first floor of an existing building block. In order to get on this floor, there is a concrete path to guide children and parents to the entrance. The same vocabulary of concrete curves is repeated in the terrace on the second floor that serves as an external playground.

This building has an open floor plan with a centralised grouping of sleeping rooms for the children. The rooms are connected with sliding doors so that the spaces can be united, grouped and divided depending on the needs of the children.

项目名称：Child Day Care Center "Nieuw Kinderland"
地点：Nieuwland 194, 1000 Brussel, Belgium
建筑师：Burobill & ZAmpone architectuur
承包商：Ibo nv, Heffen
结构工程师：Util bvba, Schaarbeek
技术工程师：Arcade&Beco, Kontich
甲方：vzw Kinderdagverblijf Lutgardisschool Elsene
艺术设计：Benoît van Innis
有效楼层面积：1,407m²
造价：EUR 2,003,226 (excl. VAT and fees)
设计时间：2013 / 竣工时间：2015
摄影师：©Filip Dujardin (courtesy of the architect)

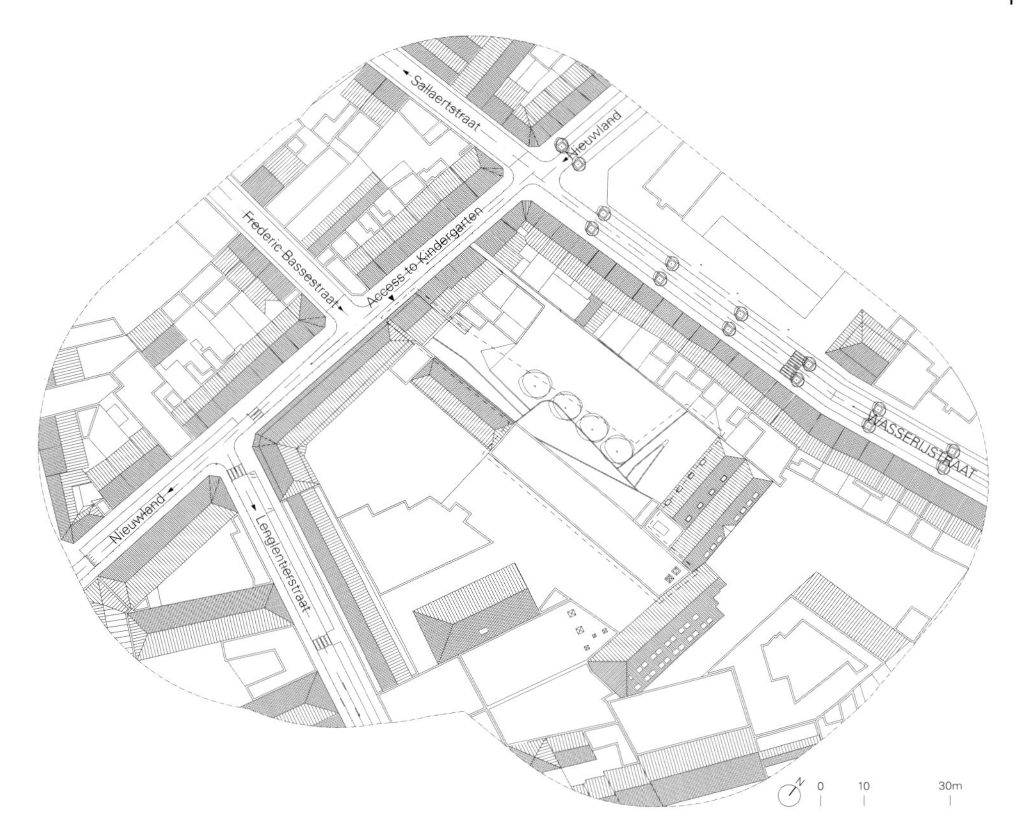

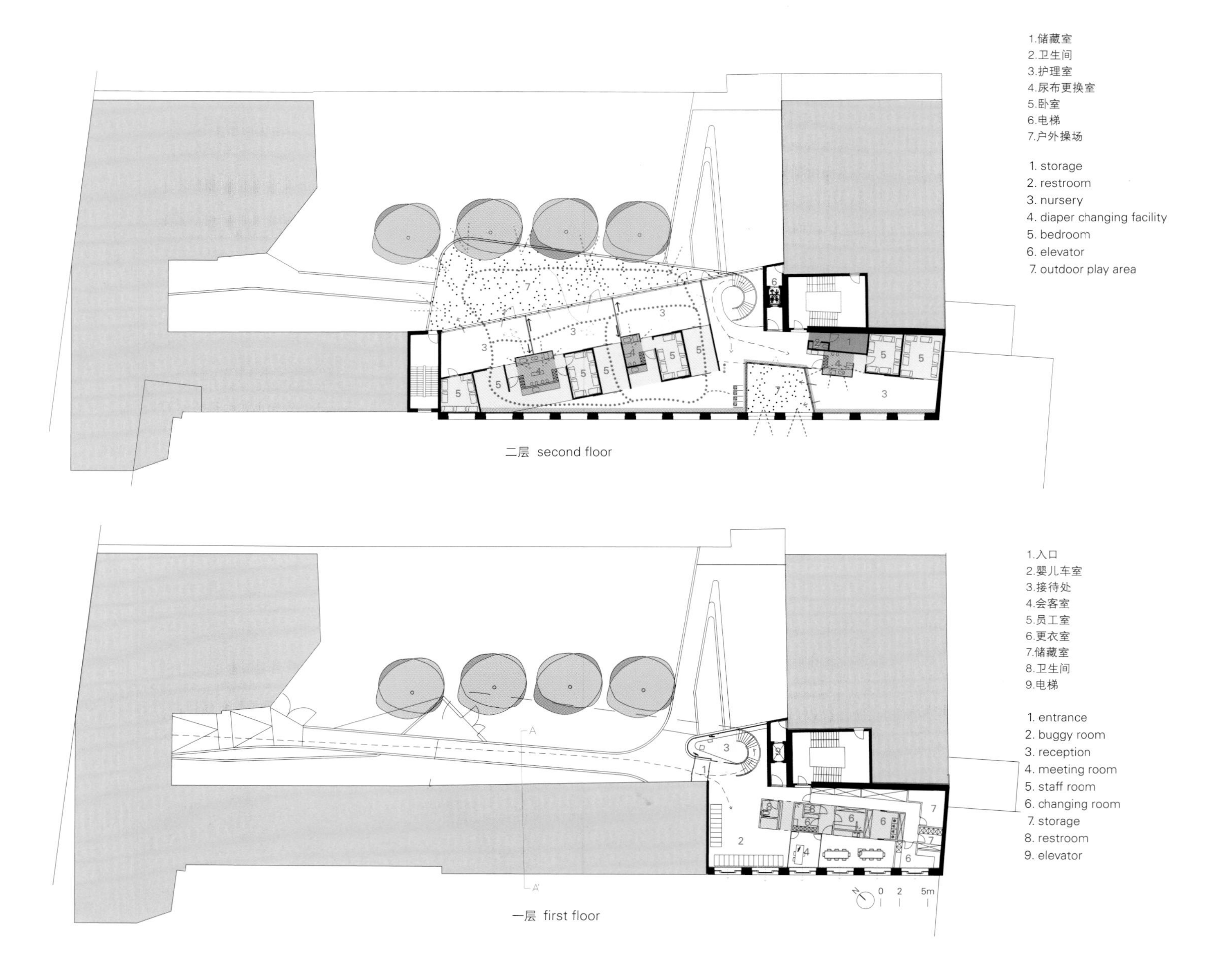

二层 second floor

一层 first floor

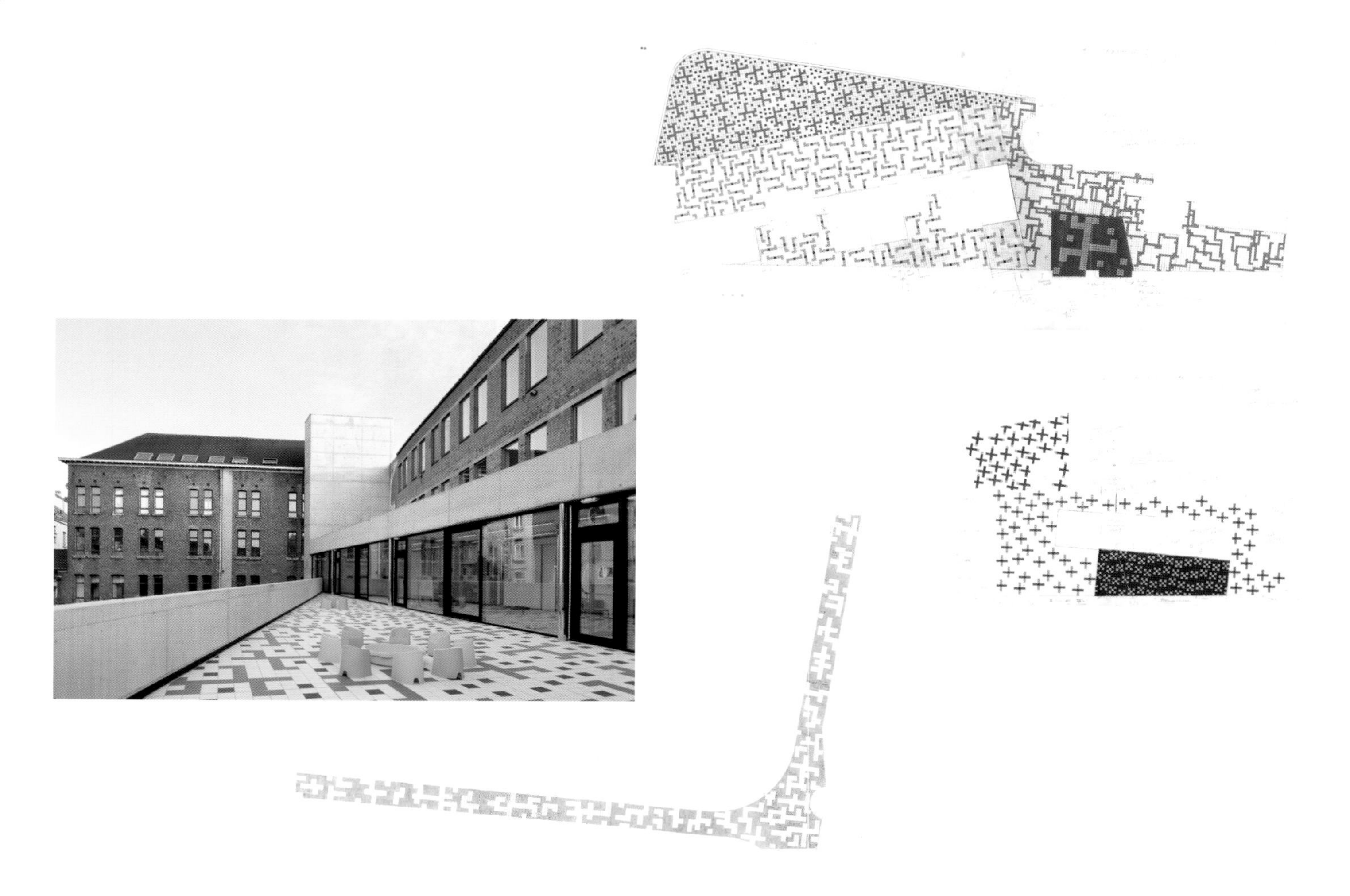

There is a corridor in the building situated on the south side of the building with large window openings. This is a buffer zone that gains heat and it can be used as the playrooms for playing with small vehicles. We call this the "playstreet". The "playstreet" gets in between seasons sufficient light through the large window openings and heat gains so that it can serve as a full-fledged play area.

All along the south facade there is a small wooden bench designed for the children to sit where they can rest or play. This gives them the opportunity to sit if they are tired of playing or they have to wait.

In order to make large openings in the existing facade to create the space for the playrooms of the children, groups of columns needed to be placed in these playrooms. We designed the columns with a mirror-finish so they integrate into the play of the children: they can see themselves in it.

The balustrades of the ramp and the outdoor play area are deliberately designed in a massive way (full concrete) in order to give the children a safe feeling. Adults however, when they are standing, can see over it and witness the outside world.

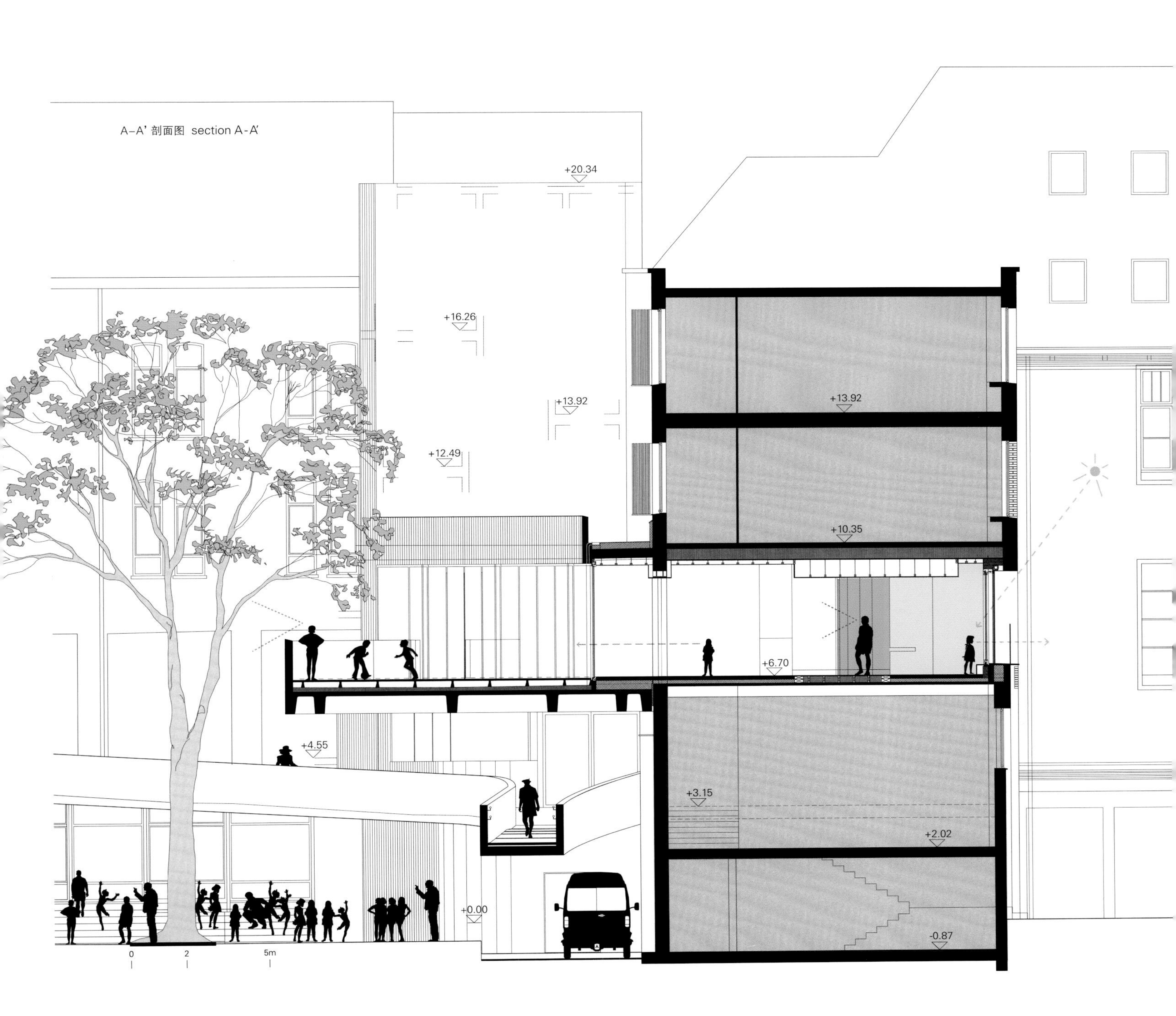

儿童空间——万花筒

A2arquitectos

这是一个改造项目，将马略卡岛的一家酒店的壁球场改造为一处儿童游乐区。该项目结合了传统的玩具——万花筒的设计理念。万花筒是经过几代人见证其视觉效果的元素，如今建筑师决定建造一个更大的万花筒，使孩子们能够与其产生互动。

可变的元素，如室外构件、自然光和孩子们的运动都会形成一系列的自然光和光反射效果，从而使空间发生变化，当孩子身处其中时，能够产生多样化的空间。

这个万花筒结构的六边形截面为 9m 长，2m 高，孩子们能走入其中，与所有的空间产生互动，从而形成万花筒效果，使孩子们望向其面对的室外公园时，便会产生一种奇特的失重感。

主入口将万花筒包裹起来，整体为粉色色调。粉色是一种令人兴奋且能激起人强烈情感的颜色，为空间营造了梦幻般的氛围。

不同大小的白色圆形图案随机出现在室内饰面中，来凸显这些空间。这些白色圆形区域还设置了不同的空间功能，用于照明、存储和隐藏空间，或者作为通向其他空间的洞口。此外，这里还设有其他空间，如浴室、楼梯、魔术道具室、潜望室以及夹层空间。这些房间对空间、颜色、反射、纹理以及氛围进行了处理，分别对应了不同的感官体验。每处场地都可分别作为独立的项目，拥有自己独特的建筑语言，但是仍

©Uschi Burger-Precht

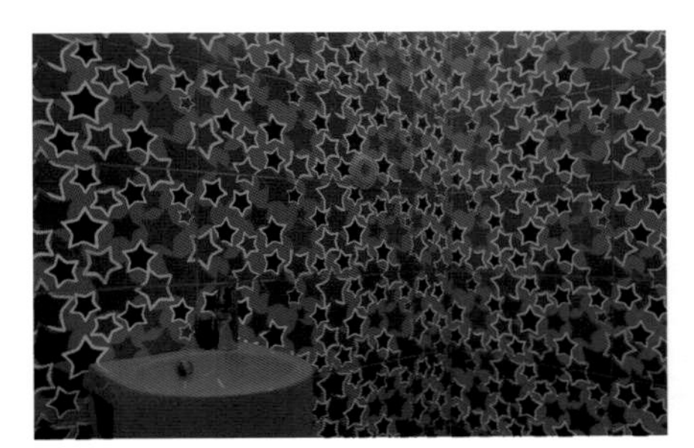

©Uschi Burger-Precht

©Uschi Burger-Precht

具有梦幻世界这一共同理念。

Space for Children – Kaleidoscope

The project is a transformation of the squash court into a leisure area for children from a hotel in Majorca, incorporating a traditional toy concept of the Kaleidoscope. It's an element that has been played by multiple generations who have observed its optical effects, and now the decisions were made to build a large scale one, so that children can interact with them.

Varieties such as external elements, natural light and the movement of children generate a set of effects of light and reflections, so that they modify the space and create multiple worlds when children are inside.

The Kaleidoscope, whose hexagonal section is of 9 meters in

length and 2 meters high, allows children to walk inside and interact with all the spaces, forming part of the own kaleidoscopic effect perceiving a strange sensation of weightlessness at the outdoor park which is oriented.

The main entrance that wraps the Kaleidoscope is characterized by its pink hue, a color which can arise the relaxing excitement but stirs powerful emotions, making this space of a dream-like atmosphere.

All of these spaces were reinforced by a random pattern of white circles of different sizes, which houses various spatial functions such as lighting, storage, slates, caches, and entries to other spaces.

The rest of rooms are additional as the bathroom, staircase, the magic room, the Periscope room, and the mezzanine. They respond to various sensory experiences playing with spaces, colors, reflections, textures and fluorescences. Treated as individual projects, each one of these places has its own language different from the rest, but keeps the common idea of the world of dreams.

项目名称：Space for children – Kaleidoscope
地点：Carretera Manacor – Porto Cristo Km. 10, Mallorca, Spain
建筑师：Juan Manzanares, Cristian Santandreu _ A2arquitectos
用地面积：31.785m^2 / 总建筑面积：90m^2
出资人：Marta Alonso, Eduardo Ramis, Andreu Ortiz
设计和施工时间：2013 / 竣工时间：2014
摄影师：
©Laura Torres Roa (courtesy of the architect) (except as noted)

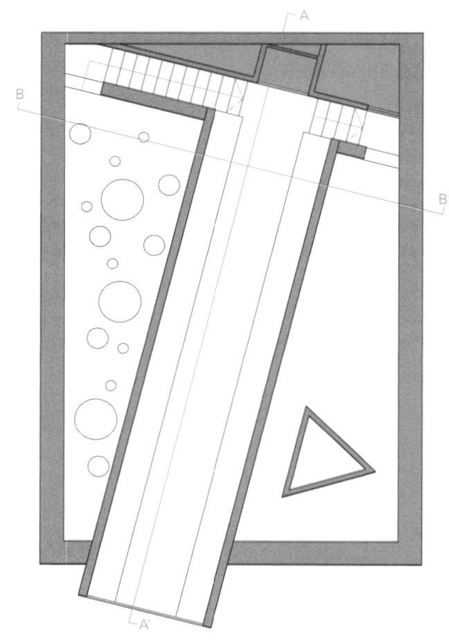

夹层 mezzanine floor

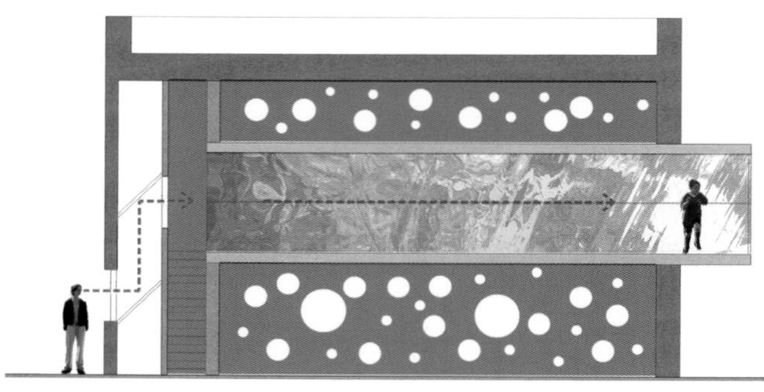
A–A' 剖面图 section A-A'

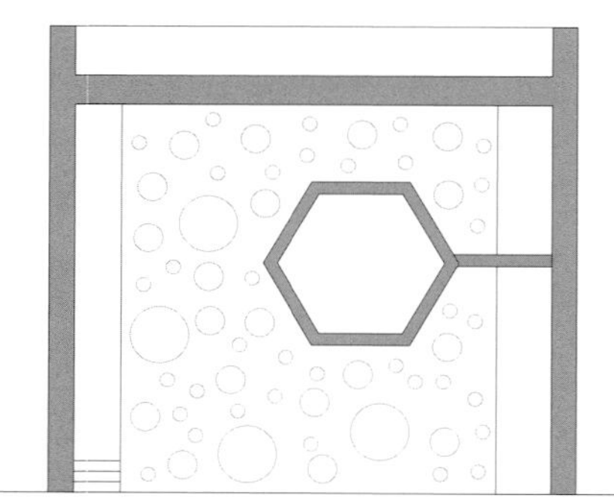
B–B' 剖面图 section B-B'

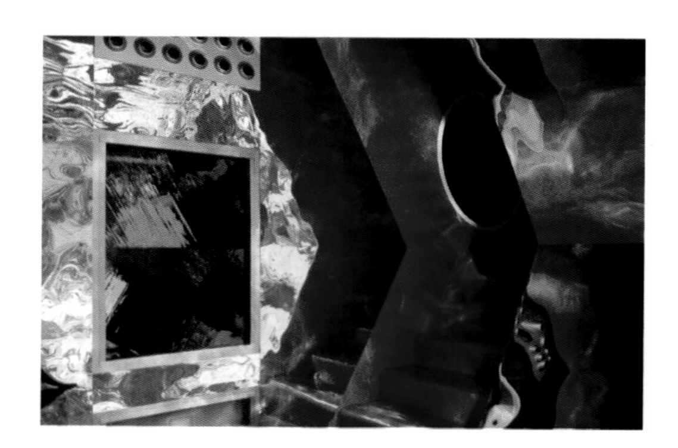

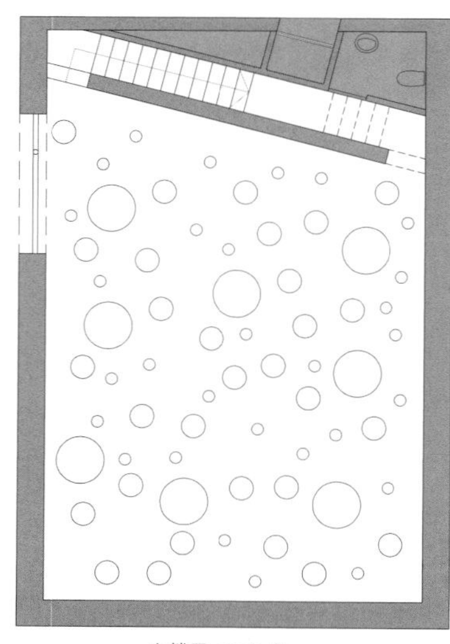

主楼层 main floor

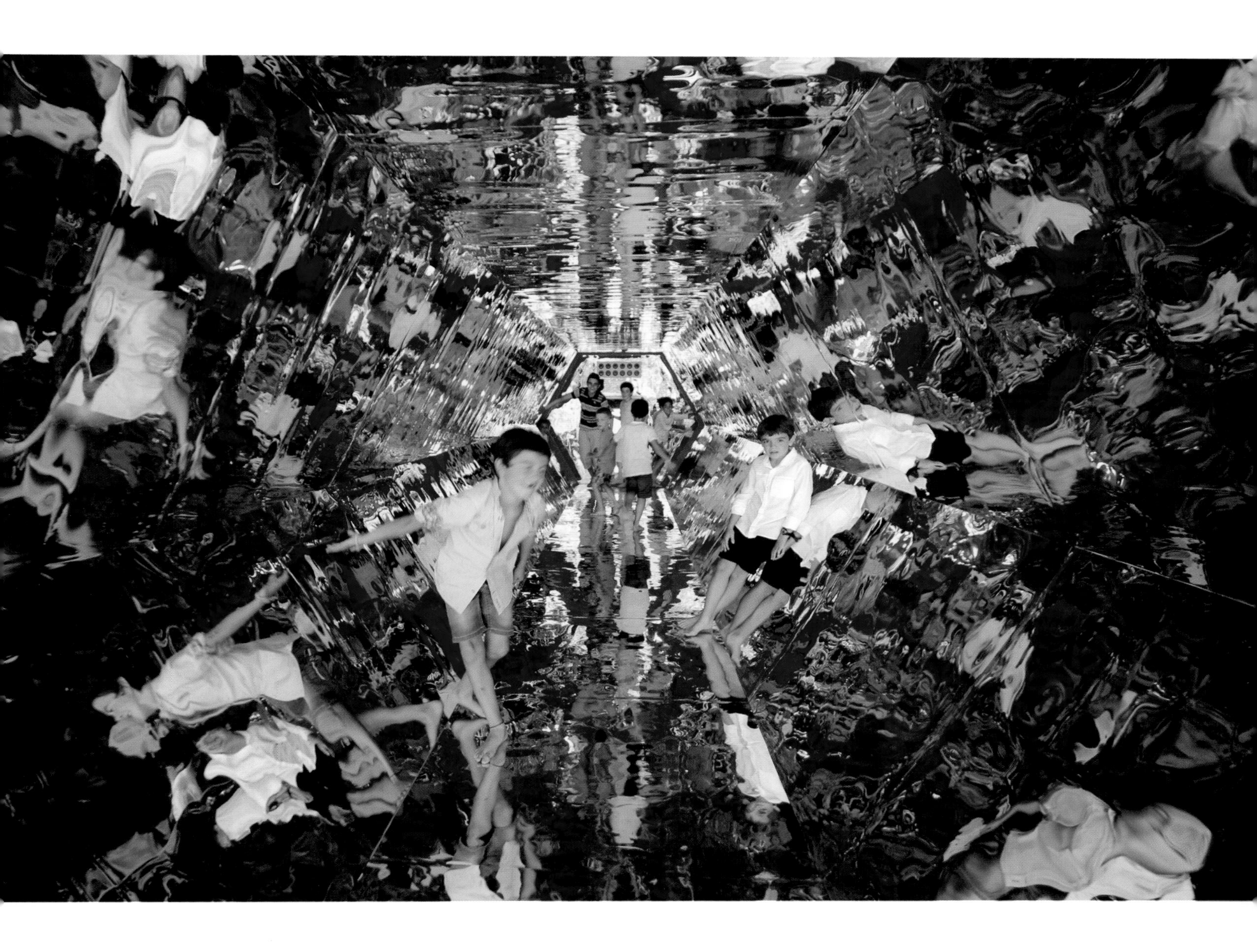

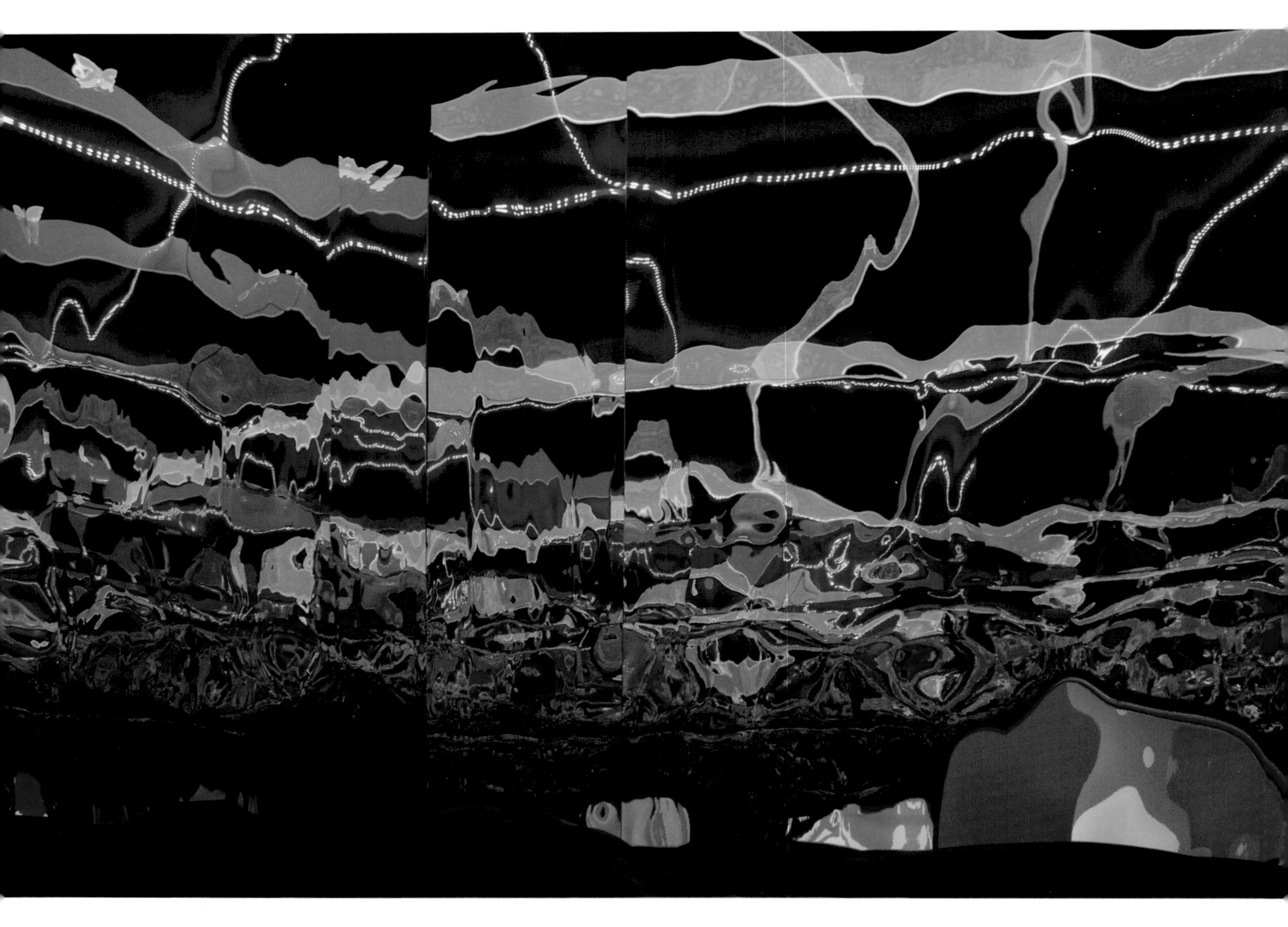

80

Ifat Finkelman + Deborah Pinto Fdeda

Ifat Finkelman completed her B.Arch with Cum Laude in 1998 and M.Sc in 2010 at the Technion, Israel Institute of Technology, Haifa, and continued her post graduate studies at the AA School, London. She is head of the Post-Industrial unit: Architecture & Production Processes at the Department of Architecture at Bezalel Academy of Art and Design.

Deborah Pinto Fdeda studied architecture at the Paris La Seine and Malaquais architecture school and received DPLG with honor in 2002. Then she graduated with the Master of Fine Art from the Bezalel Academy of Arts and Design in 2006 and worked in France, Mexico and Israel.

104

Dominique Coulon & Associés

Dominique Coulon[second from the left] was born in 1961. In 1989, created his own studio after graduation. Concepts for sustainable development, respect for historical context and responsibility towards the environment and its ecologies are amongst the main concerns for the agency. Since 2008, the agency became Dominique Coulon & Associés with Olivier Nicollas and Steve Letho Duclos as partners. Benjamin Rocchi join in as partner architect in 2014. Was nominated at the prize list of the Equerre d'Argent, Swiss Architectural Award and the Mies van der Rohe Award. Received numerous prizes including the Architizer A+ Awards and International Architecture Award from the Chicago Athenaeum Museum of architecture and Design. Currently teaches at the School of Architecture of Strasbourg.

24

Gabriel Verd Arquitectos

Gabriel Verd was born in 1975. Graduated in Architecture at the Technical Superior School of Architecture (ETSA) of Seville in 2001. Is registered architect of the Official Architects Association of Seville since 2001. Has been teaching at the International University of Andalusia, Spain, since 2005, and Architecture School of Cagliari, Italy, since 2009. Is a counsellor of the College of Architects of Seville since 2011, and was invited to join the Foundation for Contemporary Architecture in 2014.

Received International Architecture Award from the Chicago Athenaeum in 2007 and first prize "futuros" from the Superior Council of Spanish Architects' Associations (CSCAE) in 2009.

50

MAD Architects

Is a global architecture firm committed to developing futuristic, organic, technologically advanced designs that embody a contemporary interpretation of the Eastern affinity for nature. With its core design philosophy of Shanshui City – a vision for the city of the future based in the spiritual and emotional needs of residents – MAD endeavors to create a balance between humanity, the city, and the environment. Founding principal Ma Yansong[below portrait] is one of the central figures in the worldwide dialogue on the future of architecture, and has been named one of the "10 Most Creative People in Architecture" by Fast Company in 2009, and selected as a "Young Global Leader (YGL)" by World Economic Forum(Davos Forum) in 2014. Worldrenowned for works including Ordos Museum and the Absolute Towers, MAD is expanding its global presence with projects across the globe including Chicago's Lucas Museum of Narrative Art.

36

m3architecture

Was established in 1997 and is run by four Directors, Michael Banney[second from the left], Michael Christensen[first], Michael Lavery[fourth] and Ben Vielle[third]. They have won highest architecture prizes for public buildings, heritage, interiors, urban design and etc. at the RAIA Queensland Architecture Awards, AIA Queensland Architecture Awards, and Australian Timber Design Awards. Their works have been exhibited at the Venice Architecture Biennale, Australian Pavilion in 2008, 2014 and 2016.

46

Assemble + Simon Terrill

Assemble is a collective based in London who work across the fields of art, architecture and design. They began working together in 2010 and are comprised of 18 members. Assemble's working practice seeks to address the typical disconnection between the public and the process by which places are made. Assemble champion a working practice that is interdependent and collaborative, seeking to actively involve the public as both participant and collaborator in the on-going realization of the work.

Simon Terrill[right] is an Australian artist based in London. He works with photography, sculpture, video and installation. Following a BA in Sculpture and MA in Fine Art at the Victorian College of the Arts Melbourne, he lectured in Critical and Historical Studies (2005~08) and the Center for Ideas (2003~08), both at the VCA.

118

studio we architetti

Is former associate of Giraudi-Wettstein Architects and current principal of studio we architetti in Lugano, Switzerland. After an exchange program at the Harvard GSD in Cambridge, he graduated from the Swiss Federal Institute of Technology (ETH) in Zurich in 1988. Has worked for Rafael Moneo in Madrid and Manuel de Sola-Morales in Barcelona before starting his own practice, and taught at the ETH Zurich. Is member of Sachverständigenrat in St. Gallen and Redaktionskommission werk, bauen + wohnen (Zurich). Since 2015 he has been teaching at the Master of Arts in Architecture, HSLU – T&A in Lucerne.

170

144

Taller Basico de Arquitectura

Was founded in 2002 at Pamplona, Spain, by Javier Pérez Herreras and Javier Quintana. These Spanish architects represented Spain at the 8th Biennial of Architecture of Venice.
Javier Pérez Herreras is a Ph.D. in Architecture from the University of Navarra. Is currently a professor of Architectural Design at the University of Zaragoza and visiting professor of IE University in London. Javier Quintana holds a Masters in Advanced Architectural Design from Columbia University, and has a Ph.D. in Architecture from the University of Navarra. Is member of the Executive Committee of the Council on Tall Buildings and Urban Habitat from Chicago.

158

Pierre-Alain Dupraz

Was born in 1967, Geneva. Certified as architectural draughtsman in 1986. Received Architecture diploma at the Geneva School of Engineering in 1991. Collaborated with Archambault, Barthassat & Prati, Chantal Scaler and kept partnership with Christian Dupraz for 9 years. Worked as Assistant of Prof. Martin Boesch at the Architecture Institute of the University of Geneva in 2001. In 2002, established Pierre-Alain Dupraz Architecte and became a Member of the Federation of Swiss Architects. Has taught civil engineering at the Superior School of Engineering and Architecture of Fribourg (HEIA). Was a Committee member of the Federation of Swiss Architects.

128

Paul Le Quernec

Was born in Brittany, France, in 1976. Received bachelor's degree in 1994 and architect diploma at the ENSAIS, Strasbourg in 2000. Has worked with Michel Poulet, Hoog-Husson office, Chiodetti-Crupi office before establishing his own office in 2002. Taught at the INSA Strasbourg, ENSA Strasbourg and Architecture School of Nancy.

170

Aisaka Architects' Atelier

Kensuke Aisaka is First-class Registered Architect of Japan. Was born in 1973, Tokyo. Graduated from the University of Tokyo, Architecture Department in 1996. After 7 years experience at Tadao Ando Architects & Associates, established Aisaka Architects' Atelier in 2003. Has been teaching at the Toyo University Architecture Department and Aoyama Technical College. Is Member of Japan Institute of Architects and Tokyo Society of Architects & Building Engineers.

158

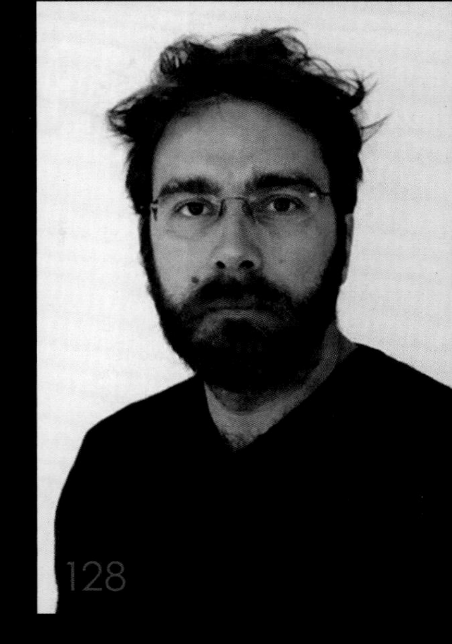

128

194

A2arquitectos

Is an architectural practice based in Madrid, Spain. Has offices in Madrid, Mallorca and Valladolid. The practice specializes in projects related to the tourist sector and has carried out numerous projects in different hotels in Spain and France. Under the direction of the architects Juan Manzanares[right] and Cristian Santandreu[left], the studio undertakes the urban planning, new building projects, renovation projects, interior design and furniture design. Received number of prizes in numerous categories at the Architizer A+ Awards, Iconic Awards and European Hotel Design Awards.

Directors, Anisara Chaktranon and Siriyot Chaiamnuay studied architecture at the Chulalongkorn University, Bangkok. Anisara Chaktranon received a Masters Interdisciplinary Design course in Interior, Industrial and Identity at the Design Academy Eindhoven. Worked at RDG planning&Design, Bangkok and Orbit Design studio. From 2007 has been a design director at Onion CO.Ltd. Siriyot Chaiamnuay received a M.Arch from AA school, London and worked for Architect 110, Bangkok and Zaha Hadid Architects, London. Has been a guest speaker at Chulalongkorn University, KMUTT(King Mongkut's University of Technology Thonburi), Chiangmai University and Kasetsart University. Was a visiting instructor at the Department of Urban and Regional Planning, Chulalongkorn University.

38

Bruno Fioretti Marquez

Piero Bruno, Josè Gutierrez Marquez, Donatella Fioretti graduated from the IUAV in 1990 and established their studio in 1995. Piero Bruno was born in Trieste, Italy in 1963. Has worked with different architectural offices in Italy and Germany. Josè Gutierrez Marquez was born in Rosario, Argentina in 1958. Has taught at different international universities including Fachhochschule Lausitz. Donatella Fioretti was born in Savona, Italy in 1962. Has worked in collaboration with Atelier Zumthor and taught at the TU Berlin. They were visiting professors at the Summer workshop of IUAV in 2004 and 2005.

70

Hibinosekkei + Youji no Shiro

Was established in 1972 and currently has office in Kanagawa, Japan. Designs various facilities mainly for children including kindergartens and nursery schools. Has designed more than 350 works in Japan and won the KIDS DESIGN AWARD in recognition for outstanding works for kids. In recent years, several foreign companies in different fields have been visiting the firm, and already started some projects in Holland and Germany. Provides comprehensive design services such as logo mark, website, uniforms, furniture, and dishes as well as architectural design.

16

Isabel Potworowski
graduated from TU Delft with a Master in Architecture, and currently works for Barcode Architects in Rotterdam. During the graduate studies, Potworowski was a member of the editorial committee and wrote several articles for the independent student journal "Pantheon". Originally from Canada, She completed her Bachelor in Architecture at McGill University in Montreal, where she was awarded the Louis Robertson book prize for the highest grade in first year history. She has also studied for one semester at the Politecnico di Milano. She has worked at ONPA Architects and Manasc Isaac Architects, both in Edmonton, Canada.

186

Burobill
Was founded in 2005 in Brussels by Peggy Geens[right], Lien Moens[center], Kristien Vanmerhaeghe[left] and has 3 coworkers. Has built large facilities for healthcare, office, commercial and mixed use, as well as small residential ones. Investigates new sustainable solutions and aims at simplicity and transparency.

186

ZAmpone architectuur
This Brussels based Belgian architectural firm was founded by Tom De Fraine[right], Bart Van Leeuw[center], Karel Petermans[left] and currently has 7 coworkers. Chooses to enrich the team, if necessary, by collaborating with external specialized studies depending on the demand of each project.

图书在版编目(CIP)数据

童趣空间 : 英汉对照 / 美国MAD建筑事务所等编 ; 周一, 丁树亭, 孙茜译. — 大连 : 大连理工大学出版社, 2017.6
(建筑立场系列丛书)
ISBN 978-7-5685-0803-2

Ⅰ. ①童… Ⅱ. ①美… ②周… ③丁… ④孙… Ⅲ. ①儿童教育－教育建筑－建筑设计－环境设计－汉、英 Ⅳ. ①TU244.1

中国版本图书馆CIP数据核字(2017)第112350号

出版发行:大连理工大学出版社
(地址:大连市软件园路80号　　邮编:116023)
印　　刷:上海锦良印刷厂
幅面尺寸:225mm×300mm
印　　张:11.75
出版时间:2017年6月第1版
印刷时间:2017年6月第1次印刷
出 版 人:金英伟
统　　筹:房　磊
责任编辑:张昕焱
封面设计:王志峰
责任校对:高　文
书　　号:978-7-5685-0803-2
定　　价:258.00元

发　行:0411-84708842
传　真:0411-84701466
E-mail:12282980@qq.com
URL: http://dutp.dlut.edu.cn

本书如有印装质量问题,请与我社发行部联系更换。